Texts in Computational Science and Engineering

25

Editors

Timothy J. Barth
Michael Griebel
David E. Keyes
Risto M. Nieminen
Dirk Roose
Tamar Schlick

More information about this series at http://www.springer.com/series/5151

Einar Smith

Introduction to the Tools of Scientific Computing

 Springer

Einar Smith
Fraunhofer Institut für Algorithmen
und Wissenschaftliches Rechnen SCAI
Sankt Augustin, Germany

Institut für Numerische Simulation
Rheinische Friedrich-Wilhelms-Universität Bonn
Bonn, Germany

ISSN 1611-0994   ISSN 2197-179X   (electronic)
Texts in Computational Science and Engineering
ISBN 978-3-030-60810-1   ISBN 978-3-030-60808-8   (eBook)
https://doi.org/10.1007/978-3-030-60808-8

Mathematics Subject Classification (2010): 97N80

Preface

This book provides an introduction to common programming tools and methods in numerical mathematics and scientific computing. In contrast to widespread standard approaches, it does not focus on a specific language, but rather aims to explain the central underlying concepts.

In general, new concepts are first introduced in the particularly user-friendly Python language and then transferred and expanded in various scientific programming environments from C / C ++, Julia and MATLAB to Maple.

This approach can best be illustrated with a recurring leitmotif: the numerical approximation of differential equations.

In the basic Python chapter, we introduce the function concept and illustrate how derivatives can be approximated using discrete difference quotients.

In the chapter on Scientific Python, we expand this idea in the approximation of ordinary differential equations. In comparison we show how the same sample problems can be solved with integrated solver operators.

We then extend the manual approximation methods to partial equations and their solution using the finite difference method. To prepare for this, the necessary linear algebra algorithms and matrix constructions, such as Poisson matrices, are developed along the way. In particular, we show that the use of sparse matrices leads to a significant acceleration of the computation.

The chapter on Symbolic Python shows how symbolic solvers can facilitate or replace the numerical solution process.

In addition, in this symbolic context we can easily explain the Galerkin method and thus pave the way for the introduction of the finite element method later.

In the C chapter we show how the sparse matrices that were already used in Python can be generated by machine-oriented, low-level programming using pointer constructions to implement linked lists.

The Julia chapter illustrates how the fundamental programming techniques discussed so far can be formulated in a promising newcomer to mathematical programming, which aims to combine the elegance of Python with the performance of C/C++.

We then show how corresponding ideas are implemented in commercial programming environments such as MATLAB and Maple.

In the Maple chapter, we expand the Galerkin method introduced in Python to the finite element method. In this way the reader is already well prepared for the general discussion in the FEniCS chapter.

In the chapters on distributed programming we show how classical sequential methods can be adopted to distributed and parallel computation, which is becoming increasingly important in the recent machine development, where multicore processors have found their way even into standard home computers. We discuss different approaches in Python, C/C++ and Julia.

The book closes with an advanced programming topic, the FEniCS project, which combines many of the previously developed techniques for the automated solution of partial differential equations using the finite element method.

The book is based on material from courses held by the author in the Department of Mathematics at the University of Bonn. Originally primarily intended for students of mathematics – at both bachelor and master level – the courses also attracted participants from other fields, including computer science, physics and geology.

The book is primarily aimed at students of mathematics and disciplines in which mathematical methods play an important role. A certain level of mathematical maturity is recommended. Technically, however, only very basic ideas from linear algebra and analysis are assumed, so that the book can also be read by anyone with a solid high-school education in mathematics who wants to understand how mathematical algorithms can be performed by digital computers. Programming experience is not required.

The book is written in such a way that it can also serve as a text for private self-study. With the exception of a few advanced examples in the MATLAB and Maple chapters, you can run all programs directly on your home computer, based on free open source programming environments.

The book can therefore also serve as a repetition and to improve the understanding of basic numerical algorithms.

Acknowledgments

The author wishes to thank Helmut Griebel and Marc Alexander Schweitzer from the Institute for Numerical Simulation at the University of Bonn for the opportunity to hold the programming courses and for their help in contacting Springer Verlag.

I would like to thank the course participants for their lively collaboration and critical comments, which have helped to transform the loose lecture notes into a comprehensive presentation. In particular, I would like to thank Angelina Steffens for proofreading the manuscript.

Very useful was also the correspondence with Chris Rackauckas, the author of the Julia differential equation package in Chapter 8, and Lisandro Dalcin, the author of the Python MPI implementation in Chapter 11.

I am also grateful for helpful suggestions from the anonymous referees.

My special thanks go to Martin Peters, Ruth Allewelt and Leonie Kunz from Springer-Verlag for their support and encouragement while preparing the book.

Bonn, September 2020

Einar Smith

Contents

Chapter 1
Introduction

> For what is mathematics other than
> logic dissolved into song?

Stanisław Lem

In public discussion an opinion is often heard that in our "digital age" computer programming is to be considered a basic culture technique, in importance comparable to reading and writing.

In the field of modern mathematics at least, there can be little doubt that this is the case.

Background

Many of the mathematical computation methods known for centuries, such as Gaussian elimination for solving systems of linear equations, receive their full power, required for instance in meteorology, only in implementations on high-performance computers. Here linear systems of equations with several hundred thousand or even millions of unknowns have to be solved in real time.

Similarly, numerical fluid mechanics requires solution of ever increasingly complex partial differential equations, which again only becomes possible with the support of powerful computers.

But even in pure mathematics the use of modern computers is becoming increasingly indispensable. The *four color conjecture*, first formulated in the midst of the nineteenth century, according to which four colors suffice to color any arbitrary map in the Euclidean plane, such that no two countries sharing a common boundary have the same color, was only proved in 1976 with the help of digital computers.

A more recent finding concerns an even older problem, the so-called *Kepler conjecture* about sphere packing in three-dimensional Euclidean space. It was formulated in the beginning of the seventeenth century. A proof based on complex computer calculations was announced in 1998. This proof was then systematically checked with

© Springer Nature Switzerland AG 2020
E. Smith, *Introduction to the Tools of Scientific Computing*, Texts in Computational Science and Engineering 25, https://doi.org/10.1007/978-3-030-60808-8_1

the strong support of computer assisted theorem provers, and finally accepted by the journal *Forum of Mathematics* in 2017.

Pitfalls and Limitations

In many cases where computers are actually used for mathematical computations, an inadequate understanding of how real numbers are processed has led to erroneous theories. The most famous example is arguably the so-called *butterfly effect* predicted in 1972: "Does the flap of a butterfly's wings in Brazil set off a tornado in Texas?"

With regard to the foundations of programming it was also found that a large number of computation methods – although theoretically correct – can no longer be carried out in practically relevant problem sizes. This is the case, for instance, for Cramer's rule for the solution of a system of linear equations, or for the problem, whether an arbitrary formula of propositional logic is satisfiable.

Within the theoretical foundations of programming and the related field of mathematical logic, it could even be proved that certain problems cannot be solved algorithmically at all. This refutes Leibniz' "Calculemus" as well as Hilbert's "In der Mathematik gibt es kein Ignorabimus" (In mathematics there is no ignorabimus).

Programming Languages

Delegation of mathematical methods to computers requires powerful communication techniques, especially suitable programming languages and problem-specific program libraries.

There are conflicting requirements for programming languages. On the one hand, they should enable a human user to formulate his or her algorithms in an intuitively understandable way. On the other hand, a language should facilitate the development of efficient machine-executable code.

The C Language

A language that is widely used also in mathematical contexts is the C language. Its strength lies mainly in system programming, so that the language constructs remain closely machine-oriented. This enables highly efficient programs, but with the disadvantage that the language often appears cryptic to beginners.

The user is forced to deal with machine-internal idiosyncrasies even in apparently straightforward problems. In particular, questions of memory allocation and corresponding pointer arithmetic are omnipresent. For instance, the memory space needed to store an $n \times m$ matrix A is requested as follows:

```
int** A = malloc(sizeof(int*) * n);
for (int i = 0; i < n; i++)
    A[i] = malloc(sizeof(int) * m);
```

It is obvious that the concern with such machine-oriented structures distracts from the real algorithmic problem in question.

Program Packages MATLAB, Maple, Mathematica

At the other extreme are powerful, comprehensive program environments that provide a wide range of data structures and algorithms in prefabricated and easy-to-use form.

For instance, in MATLAB the linear equation system $Ax = b$ is solved through

```
x = A \ b
```

The fact that the operator '\' actually performs Gaussian elimination embedded in an LU decomposition, is not visible and perhaps also not necessary for the user in practical work.

For a deeper understanding of mathematical programming it nonetheless appears advisable to start at a more fundamental level and learn the craft of translating algorithmic approaches into executable programs from scratch.

The Golden Middle Course: Python

For beginners in mathematical programming it is recommended to consider languages in which algorithmic procedures can be easily formulated in an intuitively understandable form and, if this is necessary for efficiency reasons, certain program parts can be replaced by machine-optimized code. And last but not least, the language should support the inclusion of problem-specific prefabricated program packages.

One such programming language is *Python*. Python is a well-structured, universally applicable general-purpose language. It combines remarkable expressiveness with a simple and compact language structure. Moreover, it is comparatively easy to learn.

Python is however by no means a "toy language", but widely used in practical contexts, not least in scientific computing.

To the contrary: Many concepts that can only be expressed with complicated constructions in other languages, are straightforward in Python, for instance, the definition of a functional that computes the approximation to the derivative of a function f:

```
def ddx(f):
    h = 0.000001
    def fprime(x): return (f(x+h) - f(x))/h
    return fprime
```

Organization of the Book

The book is divided into 5 parts.

Chapter 2 in Part I gives a brief introduction to the theory of computability and the principles of digital computers. This includes the presentation of a machine model closely related to real computers, and techniques to successively build a programming language that resembles basic Python. Other topics include *Church's thesis* and the notion of *Turing completeness*. With reference to "real" computers we explain the basic ideas and pitfalls of the IEEE 754 standard for the representation of floating-point numbers, a topic we return to later in the Python chapter.

Part II is the central part of the book. Here we develop various methods to transfer mathematical algorithms to the computer, using typical programming languages and environments.

Chapter 3 provides a systematic introduction to Python. The basic language is built up, explaining *variables*, *control structures*, and *collection types*, in particular *lists*. The function concept is introduced for *integer* and *real valued* functions. We emphasize that a function can also be the return value of another function. On a technical level, string formatting and file management are explained. As a more advanced paradigm, we discuss object-oriented programming, based on *class constructions*, including *inheritance*.

Chapter 4 presents the Python extensions *Numerical Python*, *Scientific Python*, and *Matplotlib* for graphical visualization. Mathematical topics include matrix algebra, in which both full and sparse variants are discussed, as well as optimization methods, non-linear equations, and ordinary differential equations. For partial differential equations we develop a program for solving Poisson equations with the *finite difference method*. We close with a *Monte Carlo method* for the computation of π.

In Chapter 5 we illustrate basic concepts in computer algebra with the *Symbolic Python* package SymPy. Topics include symbolic linear algebra and exact solutions for ordinary differential equations. At an advanced level we discuss the *Galerkin method* for the discretization of continuous operator problems, again illustrated for differential equations.

Chapter 6 develops the basics of C. In particular, we consider *arrays* and basic *pointer arithmetic*. We discuss C *structures* and their relationship to Python classes and show how pointers and structures can be combined to construct the sparse matrices encountered in Scientific Python.

Chapter 7 is concerned with C++ and its relationship to C. In particular we discuss the C++ data type *vector* and use it to significantly simplify the construction of sparse matrices. We take up object-oriented programming and class constructions from Python, where we, in particular, explain the approach of dividing class definitions into header and implementation files.

Chapter 8 introduces the Julia language that combines the ease of use of Python with the efficiency of C. We develop the basics by reformulating and expanding ideas from

Python, but also from C, here in particular the concept of compound types implemented by structures. We then discuss methods from linear algebra and differential equations. We show how one of Julia's main features, *multiple-dispatch* methods in functions, can be used to emulate class constructions in object-oriented languages.

Part III shows how the ideas developed in Part II can be implemented and worked out in typical commercial programming environments.

In Chapter 9 we discuss MATLAB, a highly integrated programming environment for many areas of scientific computing. We implement various topics that were examined in Scientific Python and Julia in MATLAB, in particular matrix algebra and ordinary differential equation solvers. Then we reformulate the solution programs for partial equations taking advantage of the MATLAB *matrix gallery*. We conclude with a brief introduction to a prefabricated graphical toolbox to support the solution of partial differential equations.

Chapter 10 is dedicated to Maple, a commercial computer algebra system for evaluating symbolic expressions. We compare Maple with Symbolic Python and summarize typical examples. We take up the Galerkin method from Chapt. 5 and expand it to a tool for solving ordinary differential equations with the *finite element method*.

Part IV contains a brief introduction to *distributed computing* based on *message passing* and *memory sharing*.

Chapter 11 explains an approach to distributed computing in Python according to the message-passing interface MPI. Examples include integral approximations and the dot product of vectors. At a more advanced level, we parallelize an iterative solution of *Laplace equations* and formulate a distributed implementation of the *conjugate gradient method* for the iterative solution of linear equation systems.

In Chapter 12 the same algorithms are revised and expanded in the more widespread MPI version for C/C++. We then explain the shared memory approach based on the C/C++ OpenMP module. As typical examples we reformulate integral approximations and the dot product of vectors. We show how function evaluation can be spread across multiple threads. We then show how message passing and shared memory approaches can be combined. Examples include an integral approximation for the calculation of π and a hybrid variant of the conjugate gradient method.

Chapter 13 introduces the proprietary approach to distributed processing in Julia, revisiting various examples from Python and C/C++.

In Chapter 14 in Part V we discuss FEniCS, a software package embedded in the Python environment for the automated solution of differential equations using the finite element method. The development takes up earlier discussions from Symbolic Python and Maple. We explain the different techniques needed to incorporate Dirichlet and Neumann boundary conditions. As a more advanced topic we discuss the *Stokes* equation, which includes both a vector-valued and a scalar-valued function. On a technical level, we illustrate different approaches to mesh generation in discrete approximations.

Part I
Background

Chapter 2
Mathematical Foundations of Programming

The use of modern digital computers is indispensable for large numerical tasks. In the weather forecast linear systems of equations in thousands, even millions of unknowns, have to be solved in a short time. Numerical fluid mechanics is based on the efficient calculation of ever increasingly complex partial differential equations.

The delegation of mathematical procedures to the machine requires suitable forms of communication, in particular programming languages and problem-specific program packages.

As mentioned earlier, there are conflicting requirements for programming languages. On the one hand, they should enable human users to formulate algorithms in an intuitively understandable manner. On the other hand, the language should help generate efficient machine-executable code. In order to develop sensible compromises, a basic knowledge of the background of modern computers can certainly be helpful.

2.1 A Simple Machine Model

We develop a simple mathematical model of a computing machine, which however already presents the essential properties even of the most advanced supercomputers.

The discussion in this section is based on the author's book *Elementare Berechenbarkeitstheorie* [12], which is unfortunately available only in German.

Definition 2.1 Our model, called *register machine*, consists of

- *registers* R1, R2, R3, ..., which can each contain an arbitrary natural number ≥ 0,
- an *arithmetic unit* that can perform
  - comparisons $x = y$, $x < y$ for natural numbers x, y, and
  - arithmetic operations $x + y$ and the *modified difference* $x - y$, which works like the normal difference, except that it cannot take values below 0.
- a *control unit* that interconnects register contents with the arithmetic unit, and takes care of command execution order.

© Springer Nature Switzerland AG 2020

E. Smith, *Introduction to the Tools of Scientific Computing*, Texts in Computational Science and Engineering 25, https://doi.org/10.1007/978-3-030-60808-8_2

Example 2.2 We show how the machine performs multiplication of two numbers x and y, stored in the registers R1 and R2.

Here is the program:

```
1:if R1 = 0 then stop
2:R3 <- R3 + R2
3:R1 <- R1 - 1
4:goto 1
```

Assume that register R3 initially contains the value 0.

The instruction in line 1 tells the control unit to load the content of register 1 into the arithmetic unit, where it is checked if the value is 0. If yes, the program execution is terminated (indicated by the stop command). Otherwise, the instruction in the following line 2 is executed.

In line 2 the contents of R3 and R2 are passed to the arithmetic unit, their sum is computed and then stored in R3, overwriting the old value (indicated by the *assignment operator* '<-').

Through the instruction in line 3 the content of R1 is reduced by 1. The *jump operation* goto in line 4 has the effect that the program is executed once more, beginning in line 1.

Hence, in every repetition round the content of R3 is increased by the content of R2. The number of passes is determined by the value initially stored in R1.

Since register R3 contains the value 0 at program start, the program run changes this value to $x \cdot y$.

Exercise Write a program that, beginning with initial values R1 = x, R2 = y, R3 = 0, performs integer division x div $y = \max\{z \mid y \cdot z \le x\}$ for $y \ne 0$, e.g., 15 div 4 = 3.

Machine Language and Higher-Level Programming Languages

Remaining in the analogy to "real" computers, the programming language introduced above could be considered as *machine language*. The language only comprises the machine's elementary commands, and commands always refer to actual fixed storage registers and line numbers.

The first step towards a higher-level language consists in using *variables*, i.e. *symbolic identifiers* instead of concrete registers, and leave it to the machine to provide suitable register addresses.

By the same token, jump targets are specified symbolically.

Example 2.3 A program, in which the multiplicands are initially stored in the variables arg1 and arg2, could then look like this:

```
1  res <- 0
2  label lbl: if arg1 = 0 then stop
3  res <- res + arg2
4  arg1 <- arg1 - 1
5  goto lbl
```

Through the command 'res <- 0' the program itself takes care of initializing the result register with the value 0. The target of the jump command is now also marked by a symbolic value lbl. It is left to the machine to determine the next command in the program-execution sequence. Note that line numbers are now no longer part of the program, but only serve to help the reader in referring to program components.

We collect the whole program in a *macro command* 'res <- arg1 * arg2', with which we then extend our language definition. The macro can now be used to denote multiplication in the form $z = x \cdot y$:

```
1  arg1 <- x
2  arg2 <- y
3  res <- arg1 * arg2
4  z <- res
```

First, the multiplicands x and y are loaded into the registers arg1 and arg2. The product is computed in line 3 and the result z returned in line 4.

Of particular importance is the introduction of *loops*, a control structure commonly employed in real programming languages. In Example 2.3 we note that the commands in lines 3 and 4 are executed repeatedly as long as the value of arg1 is greater than 0. This can be expressed in the form

```
1  while arg1 > 0 do
2       res <- res + arg2
3       arg1 <- arg1 - 1
4  end while
```

The behavior of the program is thus that the inner block consisting of lines 2 and 3 is executed as long as the loop condition (here arg1 > 0) is satisfied. Otherwise, the next instruction after the loop end (line 4) is executed. The instruction in line 3 guarantees that the value of arg1 eventually reaches 0.

Example 2.4 We illustrate the development so far with a somewhat larger example. We follow the inventor of the game of chess and the payment of his reward. Allegedly, the inventor of chess was granted a free wish by the king. His wish now was that the chessboard was to be filled with rice grains. More precisely as follows: One grain should be put on the first field, two on the second, four on the third, etc. (i.e. on the next field always twice as many grains as on the previous one). The king, astonished by this apparently modest wish, promised to grant it.

The following program computes the sum:

```
1  s <- 0; n <- 1; fv <- 1
2  while n < 65 do
3       s <- s + fv
4       n <- n + 1
5       fv <- 2 * fv
6  end while
```

In line 1 the initial values for the sum s, the field counter n, and the field content fv are set. The semicolon serves to separate the instructions. In the while loop, the number of grains in each field is added to the sum. Lines 4 and 5 compute the number for the next field. When the loop is finished, the variable s contains the number 18446744073709551615 of requested grains.

By expanding the macros, the complete "high-level" program can be translated back into machine instructions.

Church's Thesis

Now we can ask how powerful our apparently primitive computation model really is. It is rather straightforward to show that for instance the factorial function $f(n) = n!$, or the function $p(n)$ that returns each nth prime number, i.e. $p(1) = 2$, $p(2) = 3$, $p(3) = 5$ etc., are computable. The details can again be found in [12].

In fact, it is possible to formulate the following variant of the so-called *Church's thesis* that the American mathematician Alonzo Church stated for a related model in 1936:

Church's Thesis Every intuitively computable function in natural numbers is formally computable by a register-machine program.

Church's thesis is of course *not a theorem* in a mathematical sense, since a proof of the correspondence between a formal and an intuitive concept is obviously inherently impossible. The thesis is perhaps rather comparable to the findings in natural sciences, which are also accepted only because of a large number of empirical observations.

Namely, it has been discovered that a vast number of computation models all lead to the same class of computable functions. Moreover, there is not a single algorithmic method known that leads to a larger class. The thesis is today accepted by most mathematicians and computer scientists.

In order to prove that a computation model satisfies the conditions of Church's thesis, it has to be established as *Turing complete*, i.e. it has to be shown that the behavior of the so-called *universal Turing machine* can be emulated.

The Turing Machine

The Turing machine is a reference model for computability, in the sense that everything that is at all mechanically computable, is computable already with such a machine. If it can be verified that an automaton model can emulate a Turing machine, it – so to speak – inherits the complete computability power.

A Turing machine consists of a *control unit*, which is always in one of finitely many states, and an unbounded computation tape, divided into discrete fields. Each field can be empty or inscribed with a marker symbol, say 'I'. In each step exactly one field, the *work field*, is processed by a read-write head. The behavior of the machine is determined by an internal program. Depending on the state of the control unit and the inscription of the work field, the machine can change the symbol in the field, move the head to an adjacent field, and pass to a next internal state, see Fig. 2.1.

Fig. 2.1 In state q, the Turing machine reads the symbol a in the present work field (left), replaces it by a', moves the head one field to the right and shifts to the next state q' (right).

With some formal effort it is possible to show that our register-machine can in fact emulate the behavior of such a Turing machine. The interested reader is again referred to the book mentioned above.

Remark For specially interested theoretical minimalists, we mention that our register machine even remains Turing complete, if we restrict the elementary commands to:

- Test whether a register content is greater than 0
- Increment and decrement a register content by 1, where again the value may not fall below 0.

Limits of Computability

Naturally the question arises, why at all it should be necessary to contemplate the abstract theory of computability. In fact, it was for a long time assumed that every mathematical problem could eventually be solved. The search for limits of computability would hence be senseless. As mentioned in the introduction, Gottfried Wilhelm Leibniz is said to have exclaimed "Calculemus, let us compute!", indicating that every question of logical nature would eventually be solvable by mechanical procedures.

As late as in the year 1900, David Hilbert formulated this assumption as follows: "This conviction of the solubility of any mathematical problem is a powerful incentive for us to work; we have in us the constant call: There is the problem, seek the solution. You can find it by pure thinking; for there is no Ignorabimus in mathematics".

This hope was however destroyed in the following years by Kurt Gödel and Alan Turing. For the programming of computing machines, the "unsolvability of the halting problem" is of fundamental importance, namely the finding that there can be no algorithm, which can be used to check arbitrary computer programs to determine whether they terminate after a finite number of steps or not. Anyone, who has accidentally written a non-terminating while loop, would be grateful for such a control procedure.

Repercussions in Mathematics

The unsolvability results in computability theory and mathematical logic have reper-
cussions also in the realm of classical mathematics. Based on the work of Turing and
Gödel, the unsolvability of the famous *Hilbert's tenth problem* could be established in
1970.

Hilbert's tenth problem, raised in a list of centenary mathematical problems in
1900, is the challenge of providing a general algorithm that, for any given *Diophantine equation* (a polynomial equation with integer coefficients and a finite number of
unknowns) can decide whether the equation has a solution in which all unknowns
take integer values.

Remark For the specially interested: The crucial part of the undecidability proof is to
note that recursively enumerable sets coincide with certain number sets corresponding to solvable Diophantine equations.

Extended Number Ranges

The register machine can only deal with natural numbers. We briefly outline how our
model could be modified to handle extended number ranges.

For the representation of whole numbers from $\mathbb{Z}$ we could for instance exploit the
bijection $\varphi \colon \mathbb{Z} \to \mathbb{N}$, given by $\varphi(z) = 2z$ for $z \geq 0$ and $-2z - 1$ for $z < 0$. Then, of
course, the arithmetic operations must also be adapted accordingly.

Extending further, one could then represent rational numbers by pairs of integers.

Such an approach is in principle no longer possible for the real numbers $\mathbb{R}$, because of their uncountability. Here we must be content with finite *approximations*.
We do not pursue this further for the abstract model, but discuss the matter in the
context of actual digital computers in the following.

2.2 Digital Computers

In fact, our model is not that far from how real machines work. A digital computer
also consists of an *arithmetic logic unit*, a *control unit*, and a *primary storage*, i.e. storage cells in which data as well as program code are kept for immediate access.

In addition, there is also an *input/output management*, which organizes data input
and output, to the user or to other interfaces.

Modern digital computers are generally binary. This has no metaphysical reason,
but simply the one that the components are composed of switching elements, which
only know the states 'on' or 'off'. From the word origin of "digital" from the Latin word
"digitus" for finger, one would rather expect a construction based on the decimal
system.

In memory, data is kept in the form of so-called *binary words*. A memory cell can
today usually store a sequence of 0s and 1s ("bit sequences") of length 32 or 64.

Number Representation

As stated above, there is a fundamental difference in the representation of whole and real numbers.

Natural Numbers

If we interpret the contents of a 64-bit cell as a natural number in binary form, the maximum number that can be represented is $2^{64} - 1$ (that is, exactly the number of summed up rice grains in Example 2.4. Why?), for a 32-bit cell correspondingly $2^{32} - 1$.

In this sense, our register machine model is obviously an idealization. In real computers, the processing of arbitrarily large numbers cannot be an elementary operation, but has to be implemented in program code.

Whole Numbers

To include *negative numbers*, the representable range is usually shifted, in the case of 64-bit cells to the range of -2^{63} to $2^{63} - 1$, analogously for 32-bit cells to the range of -2^{31} to $2^{31} - 1$.

Real Numbers, Floating Point Numbers

Of particular importance is the representation of *real numbers*. Almost all processors and programming languages today use *floating-point numbers* in the unified binary format defined by the so-called *IEEE 754 standard*. Analogous to the scientific notation for pocket calculators, the numerical value z is split into a pair of binary numbers, *mantissa m* and *exponent e*, such that $z = m \cdot 2^e$. Here, m is a floating point number in the *normalized* representation with 1 before the separation point, such that the decimal number 5.25 (binary 101.01) is represented by $m = 1.0101, e = 2$.

The common variants with *single* and *double* precision provide 24 or 53 bits for the mantissa, such that binary numbers with a larger number of digits are rounded correspondingly.

For decimal numbers, it follows that the mantissa can only be represented exactly up to 7 or 15 digits.

Remark 2.5 Computationally, this can be established as follows: For the binary representation of a number $n \in \mathbb{N}$ one needs $\lceil \log_2(n) \rceil$ bits, for a 7-digit decimal number up to $\lceil \log_2(10^7 - 1) \rceil = 24$, while the 8 digit number $10^8 - 1$ already requires 25 bits. With 53 bits, only decimal numbers with up to 15 digits can be represented.

Another source for rounding errors is that most decimal fractional numbers cannot be represented exactly by binary numbers at all! For example, the number decimal 0.1 corresponds to an infinite periodic binary number $0.0\overline{0011}$, in normalized representation $1.\overline{1001} \cdot 2^{-4}$.

The rounding errors are then amplified within calculations. In fact, the finite precision of floating-point numbers is one of the *fundamental problems of numerical mathematics*. It affects every computation process based on real numbers.

Higher-Level Programming Languages

Computation in a real digital computer is based on commands that are actually as simple as we have assumed in our abstract model. Often it appears that the processor provides more complex instructions, but these are then usually programmed internally in the processor, which means that they are ultimately again reduced to addition and subtraction.

The reason to use a *higher-level programming language* is that it permits to formulate problem-oriented algorithmic procedures in a better human-understandable manner. Programs written in a higher language can then automatically be translated into machine-executable code.

Example 2.6 As an example we reconsider our program for calculating the rice grains in Example 2.4. Here the translation work to be performed is: replace the control structure of the while loop with jump commands, and convert the multiplication in line 5 into a sequence of elementary commands.

There are two main approaches to translating higher-level programs into machine code. In a *compiler* the entire program is translated into an independent machine program, which can then be executed directly without further reference to the higher-level source.

In an *interpreter* the program is translated line by line and each line is then executed immediately. This also means that the program cannot be run on its own but only in the dedicated interpreter environment.

The advantage of the interpreter approach is that programs can run immediately without the need for a time-consuming translation process. The disadvantage is that interpreted programs tend to run slower.

The latter can be explained by the while loop in Example 2.4. In an interpreter-based version, the multiplication macro has to be translated each time it is called, while a compiler only performs the full translation once.

Part II
Core Languages

Chapter 3
Python, the Fundamentals

Python is a well-structured, powerful programming language. There are many reasons why it is useful as an introduction to scientific computing. It combines remarkable expressiveness with a clean, simple and compact language structure. It's easy to learn. The common programming paradigms *procedural, functional* and *object-oriented* can be studied and learned in Python. However, Python is by no means a "toy language". It is one of the most widely used languages in the world, not least in the field of scientific computing.

The language was developed in the early 1990s by Guido van Rossum in the Netherlands. The name is derived from the British comedian group *Monty Python's Flying Circus*, very popular in the 1970s; and not from the snake with the same name. As a matter of fact, it is also a sketch of that group to that we owe the characterization *spam* for a flooding of serious communication with senseless disturbance. In the Python literature, even today we often find allusions to the sketches of the flying circus.

Presently, there are two coexisting Python generations, version 2 and version 3. Version 2 is still often used in practice, in particular also because some useful program libraries are not yet satisfactorily adapted to version 3.

The libraries most relevant to numerical mathematics do however work with version 3, in particular *NumPy* (for *Numerical Py*thon), *SciPy* (for *ScientificPy*thon) and *Matplotlib* for the graphical representation of functions, as well as *SymPy* for the symbolic manipulation of mathematical expressions. In the following we can therefore base our discussion on the modern variant, more precisely on Python version 3.7. For readers who prefer to work with version 2, we will generally note the relevant differences.

The `python.org` website of the *Python Software Foundation* is recommended as basic reading, in particular the tutorial `docs.python.org/3.7`. A good comprehensive reference is the 1600-page book *Learning Python* by Mark Lutz [8].

© Springer Nature Switzerland AG 2020
E. Smith, *Introduction to the Tools of Scientific Computing*, Texts in Computational
Science and Engineering 25, https://doi.org/10.1007/978-3-030-60808-8_3

3.1 Python Interpreter

Actually, we have already seen an example of a complete executable Python program: the program to compute the number of rice grains in Example 2.4 in the last chapter. To run it in Python, it is sufficient to replace a few symbols with their Python equivalents.

But what do we mean by executable? Python is a language that is normally used in an interpreter environment (in the sense explained in the last chapter). On most Unix or Linux systems, such an environment is automatically provided, or if not, easily installed.

For the home computer under Windows, Mac or Linux the Python distribution *Anaconda* is recommended. It is freely available from `anaconda.com`. Installation of Anaconda also downloads the powerful programming environment *Spyder* (an acronym for *Scientific PYthon Development EnviRonment*). Detailed documentation can be found at `spyder-ide.org`.

All examples in the following chapters were tested in Spyder running under macOS.

A First Example

In introductions to programming languages, traditionally a program is presented that writes the welcome "Hello World" to the screen.

In Python, the program consists of the single line

```python
print("Hello World")   # my first Python program
```

To run it, we launch the 'Spyder' environment from the Anaconda-Navigator. On many platforms, Spyder can also be started by entering 'spyder' in the Terminal application. We type the program code in the editor on the left side of the Spyder window. The program can then be run immediately by pressing the Run button. It can also be saved to a file for later use.

The text string is output in the Spyder console window.

Note however, that the quotation marks have been stripped off; they only serve to tell the interpreter that the *text string* between them should not be evaluated. Here we could equivalently use a pair of single quotation marks.

The data type `str` for *strings* is the first in a series of Python data types we will encounter.

The text following the '#' character to the end of the line is not processed by the interpreter. It only serves as a *comment* for the user. A block of text can also be declared as a comment by enclosing it in a pair of three quotation marks, such as `"""a comment"""` or `'''another one'''`.

Remark Note that the `print` command is presented in bold. Throughout this introductory chapter we follow the convention to mark Python *keywords*, i.e. words that belong to the core Python language, in this way.

Remark In earlier Python versions, the print command was written without parentheses in the form

```
print "Hello World"
```

Both are accepted in Python 2.7, so it seems advisable to stick to the modern form, since the form without parentheses is not usable in Python 3.

Interactive Mode

Especially for learning or testing of small programs, Python offers the convenience of *interactive programming*. The user enters the input at the *command prompt*, which in Python usually consists of three 'larger-than' characters '>>>', or, as in Spyder, has the form 'In[n]:', where n is a counter for the number of inputs within a session.

After pressing the Enter key, the result is displayed on the next line.

In interactive mode (and only here!) the print command can be omitted:

```
1 >>> 'Hello World'  # input on the command line
2 'Hello World'      # response of the interpreter
3 >>> print('Hello World')
4 Hello World
```

A subtle difference can still be seen. In line 2 the input is simply mirrored, whereas the print command peels off the quotation marks, which only served to specify the input argument as a string.

Another useful observation is that the last response of the interpreter can be recalled through input of the special symbol '_'. In line 3 above, we could equivalently write print(_). The interpreter then replaces the underline symbol '_' with 'Hello World' from line 2.

The underline symbol is convenient for instance in a sequence of computations such as

```
>>> 1 + 1  # Out: 2
>>> 3 * _  # Out: 6
```

Remark Note that we have written the outputs as comments on the same line as the input, instead of on a new line. To save vertical printing space we often follow that convention throughout.

One last tip: In most editors, the last entries can be called up with the up/down arrow keys.

3.2 Elementary Data Types

As common in modern programming languages, Python provides a basic set of elementary *data types*. We have already met the type str for the representation of *symbol strings*. In general, string processing relies on an extensive collection of handling

methods, so also in Python. In mathematical programming, however, string processing is of rather subordinate importance, so that in the following we consider the type
str only as far as necessary in mathematical contexts.

Numbers and Arithmetic

Of prominent importance in scientific computing are the *numerical data types* int
and float for the representation of whole and real number values.

Whole Numbers, Integers

For *integer numbers* from $\mathbb{Z}$, Python provides the data type int. It can represent *arbitrarily large* values:

```
1 >>> 2**100
2 1267650600228229401496703205376
```

The operator '**' in line 1 is the Python notation to express the *power function*. Here
the number 2^{100} is to be computed. In line 2 the exact value is returned.

Small numbers (according to absolute value) are processed within the arithmetical framework provided by the processor. Numbers that exceed that size are then
processed by Python itself. In theory, this is significantly slower, however in practice
barley noticeable with modern machines.

The limit is usually at $2^{63} - 1$. It can be inquired by

```
1 >>> import sys
2 >>> sys.maxsize
3 9223372036854775807
```

The program package sys is loaded in line 1, then the maximum size of numbers
that can be represented in the processor's own arithmetic is requested in line 2 and
returned in line 3.

More details regarding the package sys can again be found on the site python.org.

We'll discuss the inclusion and use of additional program packages in more detail
later.

Remark In Python 2.7 there is still a distinction between the data types int and long.
There, numbers that exceed the limit of int are distinguished by the suffix 'L' or 'l'.

Real Numbers, Floating Point Numbers

The Python data type float for the representation of real numbers is normally based
on the 64-bit precision of the IEEE 754 standard. In the following we will simply speak
of *floats*. The float numbers can be represented directly in decimal-point form, but
also in scientific notation such as 1.3e2 or 1.3E2 for 130 or 1e-6 for 0.000001.

As mentioned earlier, the finite representation of real numbers is the fundamental problem in numerical mathematics, regardless of how many bits are used. We'll come back to that later. Here we only give a simple example that confirms our theoretical observations on representation-precision in the last chapter:

```
1 >>> print('{:1.17f}'.format(0.1))
2 0.10000000000000001
```

The effect of the print command in line 1 is that the float value of 0.1, stored internally as a binary number of length 64, is returned with a precision of 17 decimal digits. We will later see in more detail how such formatted print commands are built up.

Remark For the interested reader: The number of representable decimal digits can be obtained through

```
>>> import sys
>>> sys.float_info.dig  # Out: 15
```

Observe that the number 15 confirms to our estimation in Remark 2.5 in the last chapter.

Arithmetical Operations

Python knows the usual arithmetic operators +, *, -, /, and some more, such as the power operator ** used above. A complete list can be found at python.org.

In "mixed expressions", where at least one float occurs, the result is also a float:

```
>>> 5 * (3 + 4.0)  # Out: 35.0
```

In division, the result is *always* of type float:

```
>>> 4 / 2  # Out: 2.0
```

To distinguish, the *integer division* is denoted by '//':

```
>>> 5 // 2  # integer division
2
```

The remainder of the integer division is computed with the '%' operator:

```
>>> 5 % 2  # remainder in integer division
1
```


Remark In Python 2.7 the behavior is different. There the division of integers is always interpreted as integer division with the result, say, 0 for 1/2.

If desired, the behavior of Python 3 can be enforced, like this:

```
>>> from __future__ import division
>>> 1 / 2  # Out: 0.5
```

Error Propagation in `float` Computations

As mentioned above, the fundamental problem of finite representation of real numbers can already be seen in apparently completely harmless expressions.

From the following example we can deduct a basic rule for the use of real numbers, namely *never* to test floats for equality:

```
>>> 0.1 + 0.2 == 0.3  # Out: False
```

In Python the symbol '==' denotes the equality relation (more about this later). The two sides are tested for equality, and the (obviously useless) result returned.

Let us take a closer look at what happens:

```
1 >>> print('{:1.17f}'.format(0.1 + 0.2))
2 0.30000000000000004
3 >>>  print('{:1.17f}'.format(0.3))
4 0.29999999999999999
```

The reason for the different results is that in line 2 the values of 0.1 and 0.2 are first rounded independently and then added. In contrast, the rounded value of 0.3 is displayed directly in line 4.

Example 3.1 Here is a simple example to illustrate how errors can assume astonishing orders of magnitude during a computation. We use some Python constructions that will be explained later:

```
from math import sqrt
n = 10; x = 2
for _ in range(n): x = sqrt(x)
for _ in range(n): x = x**2
print(x)
```

The program computes the sequence $a_0 = 2$, $a_{i+1} = \sqrt{a_i}$ for $i = 1, \ldots, n$, then the inverse sequence $b_n = a_n, b_{n-1} = b_n^2, \ldots b_0 = b_1^2$. Mathematically it is obvious that $b_0 = 2$. For $n = 2$ the program execution still seems close to the correct computation. For $n = 51$ the result 1.65 is already significantly off, and a further extension of the sequence to length $n = 52$ concludes with $b_0 = 1.0$.

The example can also be reproduced on any standard pocket calculator, where however the critical value for n may different.

The Types `complex` and `bool`

Python has two more built-in basic data types to represent numbers.

Complex Numbers

In Python, you can put 'j' or 'J' after a number to make it imaginary, so you can write complex literals easily:

```
>>> (1+2j)*(3-4j)   # Out: (11+2j)
```

The 'j' suffix comes from electrical engineering, where the variable 'i' is usually used for current.

The type of a complex number is `complex`. The type designator can also be used as a constructor:

```
>>> complex(2, 3)   # Out: (2+3j)
```

Boolean Operators, Comparisons

The type `bool` contains only two values `True` and `False`. As in many other programming languages, `True` and `False` are just other names for the two integers 1 and 0:

```
>>> 1 == True   # Out: True
```

Boolean expressions can be composed with the Boolean operations `and`, `or`, `not`.

Comparisons are denoted by '<' for 'less than', '<=' for 'less than or equal'.

Comparisons can also be chained, for instance:

```
>>> 1 < 2 < 3    # Out: True
>>> 1 < 2 >= 3   # Out: False
```

However, observe the notation '==' for 'is equal to' and '!=' for 'is *not* equal to'. We'll come back to that later.

3.3 Variables and Value Assignments

So far we have only been able to evaluate arithmetic expressions and return the result immediately. However, as we saw in our abstract model, *storing* values is of paramount importance. The following program computes the sum $1 + 2 + \cdots + 100$ according to the Gauss sum formula:

```
>>> n = 100   # input
>>> sm = n*(n + 1) // 2
>>> sm   # Out: 5050
```

A *variable* (here n and sm) is a designator for a memory address.

Remark Note that it is not advisable to use the identifier `sum` instead of `sm`. `sum` is a Python keyword, that denotes a function to sum up a sequence of numbers, such as `sum([3,2,1])` with a return value 6. More about sequences and the `sum` function later.

Value assignment to variables follows the

Basic Rule The expression on the right hand side of the '=' symbol is evaluated, based on the current values of the variables. The result is then stored in the variable on the left. Any previous value of the left-hand side variable is overwritten.

As customary in programming languages, the symbol '=' is used as *assignment operator*, and not as an equality relation. The reason is simply one of writing economy. Assignments are much more common than the equality relation. The latter is then represented by '=='.

In Python a variable is not of any particular data type. The following assignment sequence is perfectly possible (though not particularly meaningful):

```
>>> a = 1+2j  # a now holds a complex value,
>>> a = True  # ... now a Boolean ...,
>>> a = 'Now a string'
>>> a          # Out: 'Now a string'
```

A variable may appear on both sides of an assignment. The basic rule above applies also in this case:

```
>>> a = 1; a = a + 2  # two commands in a line, separated by ";"
3
```

Note that the input consists of two commands. This is perfectly possible. The commands, however, must be separated by a semicolon.

The semicolon also serves another purpose, namely to suppress the output of an expression:

```
1 >>> 1 + 1    # Out: 2
2 >>> 1 + 2;   # no output
```

The result '3' in line 2 is not output to the screen.

Assignments such as a = a + 2 are so common, that the standard arithmetic operators also have special versions, which can be used to update variables quickly, such as a += 2 with the meaning: "Add 2 to the current value stored in a". By the same token, commands like 'a *= b' etc. can be used.

Note that also the special symbol '_', seen in the context of interactive programming in Sect. 3.1, simply designates a variable, which the user can however only access in *reading* mode. Each time the command is executed, the interpreter assigns it the same value that it prints on the output line.

3.4 Control Structures

Program commands are executed *sequentially*, unless control flow is changed by *loops* or *branching commands*.

Loops

We first consider loops, of which there are two main types in Python, the *while* and the *for* loop.

While Loops

Here is an example of a while loop:

```
1 >>> index = 1
2 >>> while index < 3:
3 ...     print('{} times 5 is {}'.format(index, 5*index))
4 ...     index += 1
5 ...
6 1 times 5 is 5
7 2 times 5 is 10
```

The loop is initiated by the *loop head* in line 2, terminated by the colon symbol ':'.

In the following lines the ellipsis '. . .' indicates that a block of code is expected after the loop head. The *loop body* consists of the *equally* indented lines 3 and 4. Hitting the enter key *twice* terminates the block.

The loop body is repeated during program execution as long as the index variable contains a value of less than 3. The value is increased by 1 in line 4, such that the exit condition is eventually reached. Then the command that follows the indented block is executed (that is, *if* there is one).

The string between the quotation marks in line 3 contains two placeholders of the form '{}'. The suffix '.format(index, 5*index)' denotes that the placeholder will be replaced by the current values of index and 5*index (in that order). We discuss the format notation in more detail later.

For Loops

If, as here, it is already known before entering the loop, how often it will be executed, an equivalent formulation with a *for loop* might be a better choice, namely:

```
1 >>> for index in range(1,3):
2 ...     print('{} times 5 is {}'.format(index, 5*index))
```

The head of the for loop in line 1 declares that the loop body (which in the present case consists only of line 2) is to be applied successively to every element in range(1,3).

Interlude: the range Function

The basic form of the range function is range(start,stop) that produces a sequence of integers from start (inclusive) to stop (exclusive), such that in the for loop above, range(1,3) generates the indices 1, 2.

Often used is the short form range(stop) for range(0,stop).

An optional third argument can be included to specify the increment (or decrement) step, such that for instance range(1,10,2) generates the sequence 1, 3, 5, 7, 9, range(10,1,-2) the sequence 10, 8, 6, 4, 2.

When control structures are involved, immediate command line programming tends to be cumbersome. In the following we will often assume programs to be provided in form of a file.

Example 3.2 As another example of a for loop, consider a program for the *iterative* computation of the sum of the numbers 1 to 100:

```
1 n, sm = 100, 0
2 for i in range(1, n+1): sm += i
3 print('The sum is {}'.format(sm))   # Out: 5050
```

In line 1 the two assignments n = 100 and sm = 0 are are combined. Technically, we make use of a data type `tuple`, to be explained in more detail later.

Moreover, line 2 shows that a code block, which consists only of a *single line*, can also be written directly after the colon *in the same line*. We will often make use of this.

The break and continue Statements

Often it is necessary to interrupt the execution of a loop. The `break` statement terminates the entire loop containing it. Control of the program flows to the statement immediately after the body of the loop. The `continue` statement is used to skip the rest of the code inside a loop for the current iteration only.

We will see examples later.

Conditional Statements

Often the next command in a program depends on whether a condition holds or not. In Python (as in most modern programming languages) if-else-commands are used for this purpose.

In Python, the basic form is

```
if <condition>:
    <statements>
else:
    <statements>
```

where <condition> is a Boolean expression, and <statements> a block of equally indented commands.

Again, if the command block consists of a single statement, it can be written on the same line as the keywords `if` and `else`.

As a first example we consider the *Collatz problem*, a conjecture about a sequence iteratively defined as follows: Start with a positive integer *n*. Then, if the current term is even, the next term is half the current term. If the current term is odd, the next term is three times the current term plus 1. The conjecture is that the sequence will always reach 1 regardless of the value of *n*.

Example 3.3 (Collatz Problem) In Python this can be formulated as follows:

```python
1 n = 100  # input
2 while n > 1:
3     if n % 2 == 0: n //= 2   # integer division
4     else: n = 3*n + 1
5     print(n)
6 print('reached 1')
```

In the while loop the value of the variable n is modified, according to whether n is even or odd.

To test whether n is even, the operator '%' computes the *division remainder in integer division*, and the equality test '==' checks if it is 0.

If n is even, the number is replaced by its half, where we use integer division to keep the result as an integer. If n is odd, the value of n is multiplied by 3 in line 4, increased by 1 and the new value assigned to n.

The print command in line 5 provides a protocol of the computation run. It is placed outside of the if-else block, but still inside the loop block.

Finally, the print command in line 6 is executed when the loop condition no longer holds.

For the input $n = 100$ the program terminates after 25 executions of the while loop. In fact, there is no input known, for which the sequence does not finally reach 1. In spite of considerable effort, it is, however, still an open mathematical question, whether this is always the case. Paul Erdős, undoubtedly one of the most important mathematicians of the 20th century, once commented: "Mathematics is not yet ready for such problems."

Floating Point Precision Revisited: Machine Epsilon

Now that we have a precise understanding of control structures, we can return to another interesting and important observation about floating-point representation.

For a given machine the *machine epsilon* denotes the smallest positive floating point number for which

$$1 + \varepsilon > 1.$$

The following program allows to determine this value:

```python
eps, i = 1.0, 0
while 1 + eps > 1:
    eps /= 2
    i += 1
eps *= 2
print(eps); print(i)
```

The value 2.220446049250313e-16 is returned after 53 executions of the while-block. It corresponds to what is to be expected according to the IEEE 754 standard.

Remark Note the completely different result eps = 0.0 and i = 1075 in the apparently equivalent program

```
eps, i = 1.0, 0
while eps > 0:
    eps /= 2
    i += 1
print(i)
```

In Exercise 3.1 at the end of the chapter you will be asked to give an explanation of the different behaviors.

3.5 Collection Types: Lists, Tuples, Dictionaries and Sets

All Python data types considered so far are *elementary* data types. In addition to the elementary types, there are various *collection* types which, as the name suggests, collect elements of other types.

Lists

The by far most important collection type in Python is the `list` type.

The following list

```
lst = [1, 2, 3]
```

contains the numbers 1, 2, 3 in that order. To access the individual elements we use the *component* notation `lst[0]`, `lst[1]`, `lst[2]`, also known as *bracket* or *subscript* notation. Note that the index counter begins with 0. The last element of the list can be accessed by `lst[-1]`, the second to last by `lst[-2]`, etc.

The type `list` is an example of a *class* type. It comes along with a large set of access *methods*, written in the 'dot' notation. For instance, with the above list `lst`, the operation

```
lst.append(4)
```

modifies `lst`, such that it now additionally contains the number 4 at position `lst[3]`. Conversely, we can, for example, apply the operation

```
lst.remove(2)
```

to remove the entry 2 from the list, such that `lst` now consists of the numbers 1, 3, 4 in that order. Note that the *entry* 2 is removed, not the entry at *index* 2. Actually, if there are *multiple* occurrences of 2, the first one is removed. The outcome of `[2,3,2].remove(2)` is `[3,2]`.

The entries in a list can be of any type, also again lists. Here is e.g. the list representation of *Pascal's triangle* with rows 0 through 3:

```
[[1], [1, 1], [1, 2, 1], [1, 3, 3, 1]]
```

In this way we can also define matrices. However, we postpone this to the next chapter, where we will encounter more suitable data structures. In the following we refer to lists of lists simply as *tables*.

list is a Reference Type

In general, lists are used to store large data sets. In order to save memory space, in an assignment of the form b = a between lists a and b, the value of a is not copied. Instead b is only assigned the *memory address* of a, so that a and b now access the *same memory space*.

If a value is copied, we talk about *value assignment*, otherwise about *reference assignment*.

The difference can be illustrated in an everyday situation: "I have made a copy of the entrance key for you" is a value assignment. "The key is under the door mat" is a reference assignment.

The difference between the two assignment types is of central importance in most modern programming languages.

Reference assignments can lead to undesirable side effects when used without caution. The reason is that a modification of, say, a has effect also on the other one b, like so:

```
>>> a = [1]        # a is declared as a list containing the element 1
>>> b = a          # the assignment declares b as a list
>>> a.append(0)    # a now becomes [1,0]
>>> b              # but then also b has been changed to:
[1, 0]
```

Remark In fact, this difference is so important that certain data types are specifically classified as *value* types, for instance the basic number types. The others are then classified as *reference* types. When in doubt, it is easy to make an assignment as above that changes one value and then check to see if it affects the other.

A new list with the same entries is constructed with the method copy:

```
>>> a = [1]
>>> b = a.copy()
>>> a.append(0)
>>> b  # unchanged:
[1]
```

Sieve of Eratosthenes

As an example for the use of the list type, we show how to generate the set of all prime numbers below a given number n.

Example 3.4 (Sieve of Eratosthenes) We begin with a list L to store the numbers $2, 3, 4, \ldots, n$, and an initially empty list P:

```
1  n = 10                        # input upper limit
2  L = list(range(2, n+1))       # constructs a list from  range()
3  P = []                        # [] denotes the empty list
4  while L != []:
5      p = L[0]                  # the smallest number still contained in L
6      P.append(p)               # is appended to the end of P
7      for i in L:
8          if i % p == 0: L.remove(i)
9  print(P)
```

In line 2 the list L is filled with the numbers $2, ..., n$. A short explanation: The range operator produces the values one after the other, as it makes sense in a programming loop. Here they are collected together to a comprehensive whole through application of the function list.

In line 3 the list P is prepared for collecting the prime numbers. Initially it is empty. The while loop is executed as long as there are still elements left in L. The symbol '!=' is the Python notation for the relation 'not equal'. In lines 5 and 6 the smallest number p still contained in L is appended to P. (The crucial question is: Why is p prime?) Then all multiples of p (including p itself) are removed from L.

Sublists

Let 1st be the list [3,1,4,1,5]. Then 1st[1:3] denotes the *sublist* [1,4,1] from index position 1 to position 3. (Recall that indices start at 0.) 1st[:3] denotes the list [3,1,4,1] of the first elements up to position 3. Similarly 1st[2:] denotes the last elements [4,1,5] starting at index 2.

Concatenation

Let 11 be the list [3,1,4] and let 12 be [1,5]. Then the *concatenation* '11 + 12' denotes the list [3,1,4,1,5].

Remark Note that we here encounter a very common pattern in Python, the *overloading of operators*, which means that the same operator symbol denotes different operators in different contexts. In lists, the symbol '+' denotes the *concatenation operator*.

List Comprehensions

This is an extremely versatile method of creating lists. Let 1st be a list, f a function. (We formally discuss functions below.) Analogous to the mathematical notation for sets $\{f(x) \mid x \in X\}$, then [f(x) for x in 1st] is the list of function values, resulting from the successive application of f to the elements of 1st.

A finer control is achieved, when we additionally assume a filter in form of an if-condition. For instance, let `lst` be the list `[3,1,4,3]`. Then `[x**2 for x in lst if x < 4]` results in the list `[9,1,9]`, since the element 4 from `lst` does not satisfy the choice criteria.

Such a list construction is known as *list comprehension*.

Tuples

When collecting elements, the full power of lists (and the associated computation complexity) is often not required. Then the Python data type `tuple` is practical. A tuple is a sequence of elements that cannot be changed once it has been created.

A tuple is defined like a list, with the difference that now rounded parentheses are used instead of brackets:

```
t1 = (1, 2, 3, 4)
```

When declaring a tuple, the parentheses can even be left out entirely. The following definition is equivalent to the one above:

```
t2 = 1, 2, 3, 4
```

Note that we have already used tuples before, e.g. in Example 3.2.

With tuples consisting of single elements, some special care is required, however. The instruction 't3 = (1)' does not construct a tuple, but rather an `int` number, as can be verified through the query `type(t3)`. The correct definition of a single-element tuple requires a comma following the element:

```
t4 = (1,)
```

Remark In fact, in the usual mathematical notation there is often no difference between an element x and a 1-tuple that consists of this element. If X^n denotes the set of n-tuples of a set X, then for $n = 1$ the tuple $(x) \in X^1$ is normally identified with the element x itself. When programming, however, we have to make this distinction explicit.

Dictionaries

Another useful data type for the collection of elements is the type `dict` for *dictionary*, occasionally also somewhat pretentiously called *associative memory*. A dictionary consist of a set of key-value pairs. A rectangular cuboid could be defined like this:

```
c = {'width': 3, 'height': 2, 'depth': 4}
```

Keys and values are separated by a colon, the individual pairs by commas, the pairs collected within curly braces. Access to the values is then – as in lists and tuples – provided by the component operator, such that, say, `c['width']` returns the value 3. In fact, lists can be conceived as special dictionaries, where the keys simply consist of indices 0, 1, 2, ...

Sets

Python also provides a data type `set` for *sets*, which unfortunately is rarely discovered in the wild.

Sets are created by enclosing elements in curly braces:

```
>>> a = {3, 1, 2, 2, 3}
>>> a  # Out: {1, 2, 3}
```

As in mathematical sets, a Python set is an *unordered* collection of items. Sets do not record element position, nor duplicates of single elements. Hence, in particular, elements cannot be accessed by component notation such as `a[0]`.

Basic operations are:

```
>>> len(a)  # cardinality of a
3
>>> 2 in a  # test for membership
True
```

For sets the usual Boolean operations can be applied:

```
>>> b = {4, 3, 3}
>>> a | b  # union
>>> a & b  # intersection
>>> a - b  # difference
```

Also the "comprehension" syntax, seen in lists, applies to the type `set`. The set difference 'a - b' could, for instance, equivalently be defined by

```
c = {x for x in a if x not in b}
```

Note however, that the empty set cannot be specified as '{}', but instead by e.g. 'set()':

```
>>> type({})  # creates empty dictionary !
<class 'dict'>
>>> d = set()
>>> type(d)
<class 'set'>
>>> len(d)    # Out: 0, i.e. d is empty
```

3.6 Functions

An important contribution to the computing power of modern programming languages is the possibility to define *functions* that can be used repeatedly whenever needed.

Example 3.5 (Factorial) The following function `factorial` computes the product $n! = 1 \cdot 2 \cdot \ldots \cdot n$ of the first n natural numbers:

```
1  def factorial(n):
2      res = 1
3      for i in range(1, n+1): res *= i
4      return res
```

In line 1 the keyword 'def' indicates that a function definition follows, more precisely the definition of a function called factorial in *one* variable *n*.

As in the notation for loops and the if-else command, the colon ':' indicates that a block of code follows, within which the effect of the function is defined.

The function value res is computed in lines 2 and 3, and output in line 4 with the 'return' instruction.

Note in line 3 that for $n = 0$, range(1,1) is empty, such that the loop command is not executed, and the initial value 1 of res left unchanged.

The function defined in this way is tested for the argument 4:

```
print(factorial(4))   # Out: 24
```

Anonymous Functions, Lambda Expressions

For functions whose command block consists of a single expression, often a simplified definition based on a so-called *anonymous lambda expression* is preferable.

Consider for instance

```
>>> def double(x): return 2*x
```

This function can equivalently be defined by

```
>>> lambda x: 2*x
```

Note that this notation corresponds directly to the mathematical notation $x \mapsto 2x$.

A lambda expression can be applied as a function, say

```
>>> (lambda x: 2*x)(4)   # Out: 8
```

or assigned to a variable

```
>>> dbl = lambda x: 2*x
```

which can then in turn be used just as a normal function.

If a lambda expression refers to multiple variables, these are written before the colon, separated by commas, for instance:

```
lambda x, y: x + y
```

Functions as Arguments

In Python, arbitrary data types may be used as function arguments, *in particular also functions themselves*.

Example 3.6 The following function vmax computes the maximum value of a function f on $n + 1$ equidistant points $0 = 0/n, 1/n, 2/n, ..., n/n = 1$:

```
1  def vmax(f,n):
2      max_y = f(0)
3      h = 1.0/n
4      for i in range(1, n+1):
5          y = f(i*h)
6          if y > max_y: max_y = y
7      return max_y
```

In line 1 both the function f to be evaluated and the number n are declared as arguments of the function vmax(f,n). The function values are determined one after the other in the for loop and the largest y to date is stored in the variable max_y.

We test the function vmax for an example:

```
def g(x): return 4*x*(1 - x)
print(vmax(g,7))   # Out: 0.9795918367346939
```

Functions as Return Values

In Python, functions can not only be used as arguments, but also output as *return values*.

Example 3.7 We define a derivative operator ddx, that for an input function f computes an approximation to the derivative f', according to the formula

$$f'(x) \approx \frac{f(x+h) - f(x)}{h}, \quad h \to 0:$$

```
1 def ddx(f):
2     h = 1.e-6
3     def f_prime(x): return (f(x+h) - f(x)) / h
4     return f_prime
```

Line 2 defines the value of 'h', which is responsible for the accuracy of the approximation. The notation 1.e-6 stands for 0.000001 in decimal exponent representation. In line 3 a local function f_prime is defined, which is then returned as result of the function ddx in line 4.

We test the ddx operator:

```
5 def g(x): return 4*x*(1 - x)
6 print(ddx(g)(0.3))   # Out: 1.5999960000234736
7 dgdx = ddx(g)
8 print(dgdx(0.5))     # Out: -4.0000225354219765e-06
```

Line 5 defines a function g, for which we test the operator in lines 6–8.

In line 6 we print the approximate value of $g'(0.3)$.

In line 7 we assign ddx(g) to a variable dgdx, which can then be applied as a new function in line 8.

Recursion

In many cases the ability to define functions *recursively* leads to more compact and also – after getting used to – easier to understand definitions.

Example 3.8 Here is a recursive variant of the factorial function:

```
1 def factorial_rec(n):
2     if n == 0: return 1
3     return n*factorial_rec(n-1)
```

In lines 2 and 3 one could perhaps expect an if-else command. However, this is already implicitly contained in the formulation above. If the condition n == 0 holds, the command 'return 1' terminates the computation with the return value 1, so that line 3 is no longer considered.

A recursive function definition always consists of one or more base cases, for which the function produces a result directly, and one or more recursive cases, for which the program recurs (calls itself).

For simple functions, Python provides a convenient shorthand notation for such if-else cases. The factorial_rec can, for instance, equivalently be defined as

```python
def factorial_rec(n): return 1 if n == 0 else n*factorial_rec(n-1)
```

Remark Operations of the form

```python
return <expression> if <condition> else <expression>
```

as above are often referred to as *ternary*, since they involve *three* parameters. We will meet similar constructions also in various other programming languages.

Here are two more typical examples of recursive functions. We write them in shorthand notation:

Example 3.9 The *Euclidean algorithm* for the computation of the greatest common divisor of two natural numbers:

```python
def gcd(a,b): return a if b == 0 else gcd(b, a % b)
```

Example 3.10 The computation of the *Fibonacci sequence*:

```python
def fib(n): return 1 if n == 1 or n == 2 else fib(n-1) + fib(n-2)
```

Remark In the definition above we encounter a *composite test condition*. Recall that in general, conditions can be arbitrarily combined using the logical operators and, or, not.

Example 3.11 In the mathematical theory of computability (nowadays rather a discipline within computer science), the following *Ackermann function* plays a central role:

```python
def ack(x,y):
    if x == 0: return y + 1
    if y == 0: return ack(x-1, 1)
    return ack(x-1, ack(x, y-1))
```

Note that the Ackermann function is only of theoretical interest. Even with the most advanced machines it is not possible to compute it for more than a handful of small arguments. You might try an input $x = 4$, $y > 0$. In the second argument it is a bit more conciliatory. For $x = 3$ it works with $y = 8$, but then with $y = 9$ no more.

Exercise Explain the error message returned by the interpreter, observing:

```python
import sys
sys.getrecursionlimit()  # Out: 3000
```

The Quicksort Algorithm

We end our brief introduction to basic Python concepts with an example, in which both recursion and Python's sophisticated list processing play an important role; the famous Quicksort algorithm to sort a list of numbers.

Example 3.12 Here is a surprisingly concise implementation of Quicksort:

```
1  def qsort(lst):
2      if lst == []: return []
3      p = lst[0]  # pivot element
4      sml = qsort([x for x in lst[1:] if x < p])
5      grt = qsort([x for x in lst[1:] if x >= p])
6      return sml + [p] + grt
```

In line 3 the element p at index 0 is selected as pivot element.

In line 4 the list of elements < p is generated, which is then recursively sorted with qsort, such that sml contains all elements < p *sorted in ascending order*. (Recall that lst[1:] denotes the sublist of all elements beginning with the one at index position 1, i.e. all elements except the first one p.)

Dually, the list grt contains all elements >= p (except p itself) in ascending order.

In line 6 the lists sml, [p] and grt are *concatenated*, i.e. joined together to a comprehensive list.

The algorithm can be tested as follows:

```
testList = [4, 5, 7, 3, 8, 3]
print(qsort(testList))  # Out: [3, 3, 4, 5, 7, 8]
```

3.7 String Formatting

We have already repeatedly passed numbers to the print command, specifying the desired output form with the format method. However, format is still far more powerful. Let us consider two slightly more complex examples:

Example 3.13 It is often desirable to prescribe a certain number of digits for the representation of numbers, in which the values are arranged right-aligned. Moreover, it is often desirable not to print floats in their full machine accuracy.

Such fine-tuning can be controlled with the format method:

```
1  pi = 3.14159  # The constant pi is not part of core Python
2  tmplt = '|{:2d}| times pi is |{:6.2f}|'
3  for n in [3, 4, 10, 100]:
4      outpt = tmplt.format(n, n*pi)
5      print(outpt)
```

The variable tmplt is of type str. It has two placeholders, which are filled by applying the method format. The first, {:2d}, provides 2 digits for an integer number. (The 'd' stands for *decimal integer*.) The second, {:5.2f}, can contain a 5-character float (including decimal point), where 2 places are reserved for the digits after the decimal point.

The meaning of the colon symbol is illustrated in Example 3.15 below.

In line 4 the formatted output strings for the different values of n are generated, and stored in the variable 'outpt'. The outputs, printed in line 5, then look like this:

```
6  | 3| times pi is |  9.42|
7  | 4| times pi is |12.57|
8  |10| times pi is |31.42|
9  |100| times pi is |314.16|
```

Line 9 illustrates what happens, when the reserved number of digits is exceeded.

Example 3.14 The following example shows how assignment order can be controlled:

```
>>> tmplt = '{0} + {1} = {1} + {0}'
>>> print(tmplt.format(3,4))
3 + 4 = 4 + 3
```

Example 3.15 The two types of format specification can also be combined. Here the colon symbol serves as a separator:

```
>>> print('|{1:2d}|{0:4d}|'.format(3,4))
| 4|   3|
```

3.8 Writing and Reading Files

Also in mathematical programming, reading data from – and saving to – files is of course essential. We illustrate how to do this in Python.

We first consider the handling of strings of type str.

Writing Strings

Writing to a file looks e.g. like this:

```
1  wf = open('parrot.txt', 'w')
2  wf.write('The parrot is dead!\n')
3  wf.write('It is only resting.')
4  wf.close()
```

In line 1 the open command opens a file called parrot.txt (first *creating* it, if it does not already exist). The second argument 'w' in open states that the file is opened for *writing*. The file is then linked to the file-handler variable wf, so that it can be accessed from the program.

The write method in line 2 outputs the string 'The parrot is dead!\n' to the file. Here '\n' is interpreted as a *control character*, which causes a line break.

In line 3 another text line is appended to the end of the file. In line 4 access to the file is terminated, the connection to the file handler wf cut off.

Interlude: Access Paths

The open command in line 1 refers to a specific "working directory" where the file is to be stored, by default typically the owner's home directory. If another directory is desired, an explicit path must be given, such as '/Users/myName/Desktop/parrot.txt', if it is to be stored on user myName's desktop.

A convenient way to deal with file paths is to use *string concatenation*. For example with

```
>>> dir_path = '/Users/myName/Desktop/'
```

we could access the file on the desktop by the path "dir_path + 'parrot.txt'".

For simplicity we continue to assume all files to be stored in the user's home directory, such that no explicit access path is required.

Reading

Reading from the file then looks like this:

```
1 rf = open('parrot.txt', 'r')
2 fstr = rf.read()
3 print(fstr)
```

In line 1 the parameter 'r' indicates that the file is to be accessed in *reading* mode. In line 2 the entire file content is assigned to the variable fstr as a string, including the tag '\n' denoting the line break. Therefore the print command in line 3 causes the text to be output in two lines.

Numbers

In fact, the case above is already the general one. *Only strings* can be stored in files, no numbers. Number descriptions must therefore be wrapped in character strings before being written to the file.

This is also illustrated with a simple example:

Example 3.16 We consider a number table tbl and convert it to a character string tblstrg as follows:

```
1 tbl = [[1, 2, 3], [4, 5, 6]]
2 tblstrg = ''
3 for r in tbl:
4     for num in r: tblstrg += ' {}'.format(num)
5     tblstrg += '\n'
```

Line 4 is central. For each number num in row r of the table, a space symbol followed by a string representation of the number is appended to tblstrg.

The string representation of the row is then terminated by the line-break character '\n' in line 5.

After the first run of the outer for loop, `tblstrg` looks like this: `'1 2 3 \n'`, after the second (and last) like this: `'1 2 3 \n 4 5 6 \n'`.

The inverse conversion of the string `tblstrg` into a table `in_tbl` is then as follows:

```
1  row_lst = tblstrg.split('\n')
2  in_tbl = []
3  for r in row_lst:
4      nums  = [int(c) for c in r.split()]
5      if nums == []: break
6      in_tbl.append(nums)
```

In line 1 the string `tblstrg` is divided at the '\n'-marks, and the substrings collected into a list `row_lst`, which then looks like this: `[' 1 2 3',' 4 5 6','']`. The last empty string results from the fact that we added a line break at the end of the line.

In the following for loop, each `str` component r in `row_lst` is converted to a list of numbers and inserted into the table `in_tbl`. Here line 4 is central. First, r is split up at the empty-space positions by `split()`, then each component c is converted to an *integer* number by the function `int()`, and finally collected into a new list `nums`, say, `[1, 2, 3]`.

In line 6 the list `nums` is appended to the end of `in_tbl`.

Line 5 captures the empty-string case above. The command `break` ends the loop execution and program control jumps to the next statement after the loop.

Remark Recall that the `break` command terminates the loop's *entire* execution, whereas in contrast, a `continue` command causes the loop to skip the remainder of its body in the current iteration round and immediately start the next one.

Binary Files and `pickle`

The approach to encoding data in strings has a great advantage. The resulting text files can be read also from other programs, and then processed further.

In many cases, however, you only want to back up the data for your own needs or pass it on to other Python users. Then there are solutions, which leave the coding and decoding to Python itself. One of the standard mechanisms is implemented in the Python module `pickle`.

We briefly illustrate how it works:

Example 3.17 First we have to provide the functions `dump` and `load`. They do not belong to the core Python language, but are contained in an external library `pickle`. We can add them to our Python environment by

```
1  from pickle import dump, load
```

We'll come back to the library import mechanism in general in much more detail later. Here it is sufficient to note that we can now use `dump` and `load` like any other built-in Python function.

As an example, we store the following table using `pickle` encoding:

```
2 tbl = [[1, 2, 3], [4, 5, 6]]  # input
```

We open the file `datafile.pkl` (creating it, if it does not yet exist), where the table is to be stored:

```
3 fwb = open('datafile.pkl', 'wb')
```

Note that the second argument in `open` is now `'wb'`. This means that the file is to be written with *binary data*, no longer with normal readable text.

The `dump` command in line 4 causes the table `tbl` to be encoded and then written to the file:

```
4 dump(tbl, fwb)  # write
5 fwb.close()
```

To regain the content we open the file in 'read binary' mode, and use `load` to assign the decoded content to the variable `in_tbl`:

```
6 frb = open('datafile.pkl', 'rb')
7 in_tbl = load(frb)  # read
```

Finally, we check if the original input has survived the procedure unharmed:

```
8 print(in_tbl)
```

3.9 Object-Oriented Programming and Classes

An extremely versatile idea in modern programming is to collect data types together with their application tools into functional units. To this end, Python, like many other languages today, follows the paradigm of *object-oriented* programming, and in particular implements the concept of *class constructions*.

A *class* is essentially a data type combined with various access *methods*, i.e. special functions that only apply to objects of this class.

We have in fact already made heavy use of one prominent example. The data type `list` is a class type. We have also encountered some `list` methods, such as `append` and `remove`. Typical for class methods is that their application is denoted in the so-called *dot notation*, such as `lst.append(3)` or `lst.remove(2)`.

Example 3.18 As an introductory example for object-oriented programming, we develop a class for the representation and computation of *fractions*.

It seems natural to represent a fraction like 3/4 as a pair $(3, 4)$ of numerator and denominator. The data part of a simple class `Fraction` could then look like this:

```
1 class Fraction:
2     def __init__(self, num, den):  # initialization
3         self.num = num             # internal storage
4         self.den = den
```

In a sense, the class is a blueprint for the construction of objects that are to behave as fractions. An actual *object* (in object-oriented talk: *instance* of the class) is then constructed with an instruction of the form a = Fraction(3,4), assigning it to the variable a.

In more detail the process is as follow: First, the *initializing* function __init__ in lines 2–4 allocates memory for two *internal* variables self.num and self.den, and assigns the *external* input values of num and den (here the numbers 3 and 4) to these variables. This memory space is then associated with the variable named a.

The use of the same designators num and den for both internal and external references is unproblematic and in fact quite customary.

Now the inner components can be accessed via a.num and a.den, for instance:

```
>>> a.num, a.den   # Out: (3, 4)
```

Actually we could also access the components in writing mode, and for instance change the numerator of a to 5 with the assignment a.num = 5. We will not need that here, however.

For the processing of Fraction objects we add two *methods*. In object-oriented speech, a method is simply a *function* defined *within* a class.

First we define a method to add fractions. The following lines are indented to indicate that they still belong to the code block initiated in line 1:

```
5    def add(self, b):              # fraction addition
6        return Fraction(self.num*b.den + b.num*self.den,
7                        self.den*b.den)
```

The add method serves to add a second fraction b to an already existing one a. It is applied like this:

```
>>> b = Fraction(1,2)
>>> c = a.add(b)
>>> c.num, c.den   # Out: (10, 8)
```

Observe that the method application appears in the so-called *dot notation*, where the function is written as postfix in only *one* variable, instead of – as one could perhaps expect – in the form add(a,b). However, what happens is that the method belonging to the distinguished object a is applied, for which then the other object b becomes the function argument.

If we now call to mind that the placeholder self in lines 6 and 7 stands for the object that applies the method, the effect of add should be clear.

The following method isEqualTo, as the name suggests, serves to test whether two fractions are equal. It is applied in the form a.isEqualTo(b), and returns one of the Boolean values True or False.

```
8    def isEqualTo(self, b)  # equality between fractions
9        return True if self.num*b.den == self.den*b.num else False
```

Note that in line 9 we use the shorthand notation introduced in the section on recursive functions.

Example (with a, b, c as defined above):

```
>>> d = b.add(a)
>>> c.isEqualTo(d)   # Out: True
```

From a mathematical point of view, it is not entirely satisfactory that the arguments a and b are not treated symmetrically.

In certain cases, however, Python provides special identifiers that permit more familiar expressions. If we replace 'add' by '__add__' in line 5 in the class definition above, we can write fraction addition in the usual form 'a + b' instead of a.add(b). Note that this is again an example of operator overloading. Depending on context, the symbol '+' can denote different operations.

Similarly, in line 8 the designator isEqualTo can be replaced by '__eq__', such that the equality test for two fractions a and b can be performed as 'a == b'.

Remark Observe that we follow the usual convention to use initial capital letters in class names, and lowercase letters for methods.

Exercise In Sect. 3.5 we noted the distinction between *value* and *reference* types. To which do class objects belong?

Polynomials

As a somewhat larger, and in numerical mathematics obviously more useful example, we develop a class Polynomial for the representation of polynomials over the real numbers $\mathbb{R}$.

Example 3.19 We choose to represent polynomials by *lists* of floats, in *ascending* order of the powers of x, e.g. the polynomial $p(x) = 3x^2 + 2x + 1$ by [1, 2, 3]. The class instantiation is to be performed in the form p = Polynomial([1,2,3]).

In the basic version, the class is to support *application* to arguments, such as, say, $p(4) = 57$. We will then include *addition* $p + q$ and *multiplication* $p \cdot q$, and testing for equality $p \overset{?}{=} q$.

The basic class Polynomial then looks like this:

```
1  class Polynomial:
2      def __init__(self, coeff):   # data initialization
3          self.coeff = coeff        # internal storage
4      def __call__(self, x):        # method polynomial application
5          s = 0
6          for i in range(len(self.coeff)): s += self.coeff[i]*x**i
7          return s
```

In the initialization method '__init__', the coefficient list from the input is assigned to the internal variable self.coeff.

The method '__call__' computes the function value for a polynomial p applied to an argument x, expecting an input in the form p(x). The function len in line 6 returns the length of a list, i.e. the numbers of elements contained in it.

To illustrate the development so far:

```
>>> p = Polynomial([1, 2, 3])
>>> p(4)   # Out: 57
```

Next, we include a method for *polynomial addition*:

```
8    def __add__(self, q):  # method polynomial addition
9        lst = []
10       if len(self.coeff) > len(q.coeff):
11           lst += self.coeff
12           for i in range(len(q.coeff)): lst[i] += q.coeff[i]
13       else:
14           lst += q.coeff
15           for i in range(len(self.coeff)): lst[i] += self.coeff[i]
16       return Polynomial(lst)
```

The method '__add__' serves to add polynomials p, q in the user-friendly form p+q (similar to the addition in the class Fraction). In line 9 an empty list lst is generated. The sums of the corresponding coefficients are then appended to the list one after the other.

Here we have to take into account that, in general, the coefficient lists are of different length. In the if-else block, first the longer one of the two is appended to the empty list lst in lines 11 or 14. In the following for loop the components of the shorter one are then added. The return value is a new polynomial with the added-up coefficients.

Remark One could be tempted to use an assignment of the form 'lst =' instead of the concatenation 'lst +=' in lines 12 and 15. However, as we have already seen in Sect. 3.4, this could possibly have undesired side effects.

We come to multiplication:

```
17   def __mul__(self, q):  # method polynomial multiplication
18       d1, d2 = len(self.coeff), len(q.coeff)
19       lst = [0 for i in range(d1 + d2 - 1)]
20       for i in range(d1):
21           for j in range(d2): lst[i+j] += self.coeff[i]*q.coeff[j]
22       return Polynomial(lst)
```

The method '__mul__' computes the coefficients of the product polynomial according to the standard procedure for polynomial multiplication.

In line 19 a list of the required length is defined and filled with the placeholder values 0. This is an initialization method that occurs so often, that for instance NumPy, discussed in the next chapter, offers a special function zeros exactly for this purpose.

The analysis of the rest of the method's operation is left to the reader.

Finally we include an equality test 'p == q' for polynomials. Intuitively, two polynomials are equal if their coefficient lists are equal. However, we have often stressed that we should not rely on equality tests between float numbers.

We interpret equality as "indistinguishable within machine precision", and arrive at:

```
23      def __eq__(self, q):
24          d = len(self.coeff)
25          if d != len(q.coeff): return False
26          for i in range(d):
27              if abs(self.coeff[i] - q.coeff[i]) > 1.e-14: return False
28          return True
```

Line 25 states that polynomials with coefficient lists of different length cannot be equal. Otherwise, the coefficient pairs are compared, and if they at some point differ by more than 1.e-14, the polynomials are considered *not* equal.

Only if all comparisons remain within the tolerance, we return the verdict `True`.

To illustrate the methods in the class `Polynomial`, we test the *associative law*:

```
>>> p = Polynomial([1,2])
>>> q = Polynomial([3,4,5])
>>> r = Polynomial([6,7,8,9])
>>> p*(q + r) == p*q + p*r   # Out: True
```

Inheritance

There is rarely a discussion about object-oriented programming without anyone mentioning "inheritance". To illustrate the idea, we discuss a class `Parabola` for polynomials of degree 2. It should include addition, multiplication, but in particular also a method to find roots according to the classic quadratic formula.

We could start the definition from scratch, but a more economic approach is to take advantage of the development already included in the class `Polynomial`, since parabolas are of course special polynomials.

We follow the latter and arrive at the idea of *subclassing* and *inheritance*.

Our class is introduced as follows:

```
1  class Parabola(Polynomial):
```

This declares `Parabola` as a class, but more specifically as a *subclass* of Polynomial, which basically means that we have full access to the *data structures* and *methods* defined in `Polynomial`.

However, we do specify that only polynomials of degree 2 are accepted as parabolas. To this end we formulate the following `init` function:

```
2      def __init__(self, coefficients):
3          if len(coefficients) != 3:
4              print('no parabola')
5              return
6          else: super().__init__(coefficients)
```

The init function rejects input of lists that do not consist precisely of three coefficients, required for parabolas.

Otherwise, in line 6 the input is handed over to the `init` function of the *superclass* `Polynomial`, such that then as before the coefficients are imported into the internal variable `coeff`.

Let's check what we have so far:

```
>>> p = Parabola([1, 2])        # Out: no parabola
>>> p = Parabola([1, 2, 3])
>>> p(4)                        # Out: 57
```

Also the other `Polynomial` methods can be used for parabolas.

Let's include a new method `roots` to compute the roots of a parabola. Actually we defer the real implementation to the next chapter, where we introduce the necessary mathematical apparatus. Here we only define a placeholder:

```
7    def roots(self):
8        print('To be implemented')
9        return
```

Let's try it:

```
>>> p = Parabola([1, 2, 3])
>>> p.roots()   # Out: To be implemented
>>> (p*p).roots()
AttributeError: 'Polynomial' object has no attribute 'roots'
```

This last response is acceptable, since the polynomial p*p is in fact no parabola, where the `roots` method would make sense.

However, we receive the same response for 'p + p', which *is* a parabola. The remedy is to modify the `Polynomial` method `add` to one where the return value is of type `Parabola`:

```
10   def __add__(self, q):
11       lst = [0, 0, 0]
12       for i in range(3): lst[i] += self.coeff[i] + q.coeff[i]
13       return Parabola(lst)
```

We try again:

```
>>> p = Parabola([1, 2, 3])
>>> (p + p).roots()   # Out: To be implemented
```

Now it works. What has happened, is that the new `Parabola` method '_add_' *overrides* the one in the superclass `Polynomial`.

3.10 Exercises

Machine Epsilon

Exercise 3.1 Consider the two while loops discussed in connection with the machine epsilon in Sect. 3.4.

Give a detailed explanation of the different behaviors. To this end, it might be useful to consult e.g. the article en.wikipedia.org/wiki/IEEE_754 on the numerical standard in floating-point computation, and/or follow the program execution by inserting print commands.

Polynomial Class

Exercise 3.2 Extend the class `Polynomial`, discussed in Sect. 3.9, by methods to compute the derivative and antiderivative of polynomials, and illustrate them for $p(x) = 3x^2 + 2x + 1$.

Linear Algebra

Exercise 3.3 Define a basic `Matrix` class. It should include matrix multiplication in the form $A * B$, a method to print a matrix to the screen in the usual form as a vertical sequence of rows, such as e.g.:

```
1. 2. 3.
4. 5. 6.
```

and a test for equality using the infix-operator '=='.

Give an example to illustrate that multiplication is associative.

The solutions to the following exercises can either be formulated as "stand-alone" programs or, preferably, included as methods in the above matrix class, or in a subclass.

Exercise 3.4 Compute the solution of the linear equation system $Ax = b$, where

$$A = \begin{pmatrix} 1 & 2 & 3 \\ 4 & 5 & 6 \\ 7 & 8 & 8 \end{pmatrix}, \quad b = \begin{pmatrix} 2 \\ 3 \\ 5 \end{pmatrix}.$$

Exercise 3.5 Compute the determinant of the matrix

$$A = \begin{pmatrix} 3 & -1 & 2 \\ -3 & 4 & -1 \\ -6 & 5 & -2 \end{pmatrix}.$$

Exercise 3.6 Compute the inverse of the matrix A in Exercise 3.5.

Numerical Integration with the Trapezoidal Rule

A common technique for approximating the integral of real-valued functions f over an interval $[a, b]$ is based on the following (simple) *trapezoidal rule*:

$$\int_a^b f(x)\, dx \approx (b - a) \frac{f(a) + f(b)}{2}.$$

If f is twice continuously differentiable in $[a, b]$, then the following estimate for the deviation $E(f, a, b)$ from the exact value holds:

$$(*) \qquad |E(f,a,b)| \le \frac{(b-a)^3}{12} \max_{a \le x \le b} |f''(x)|.$$

To better approximate the integral, we subdivide the interval $[a,b]$ into n adjacent equal-size subintervals of length $h = \frac{b-a}{n}$. In each subinterval we apply the simple trapezoidal rule, and then add the resulting approximations. This gives the *composite trapezoidal rule*.

Note Some functions and constants needed for the exercises below can be found in the package `math`. They can be accessed, for instance, through

```
from math import sin, pi
```

Exercise 3.7

(1) Define a Python function

```
trapeze(f,a,b,n)
```

to compute the integral of a function f over an interval $[a,b]$ according to the composite trapezoidal rule for n subintervals. Avoid repeated computations of the individual function values $f(x)$.

(2) Use the `trapeze` function to approximate the integral $I := \int_0^\pi \sin(x)\,dx$, such that the result becomes correct to 6 decimal places. To do this, use the exact value $I = 2$, which results from the evaluation of the antiderivative $-\cos$ of sin.

Exercise 3.8 In the refinement of the approximation it is a good idea to consider n-values of the form 2^m, i.e. powers of 2, since you can then reuse the already calculated values from the previous iteration round.

(1) Compute the approximation of $\int_0^\pi \sin(x)\,dx$ for $n = 2^m$, $m = 1, \dots, 10$. Store the results in a list. Print the result of the individual calculations as well as the deviations from the exact value.

(2) What assumption about convergence speed can be deducted?

(3) Check your assumption for the integral

$$\int_0^2 f(x)\,dx \quad \text{for} \quad f(x) := 3^{3x-1}.$$

In the last two exercises, the integral could be calculated directly by considering the antiderivative. Of course, numerical integration is of particular interest if the integral cannot be expressed in a closed representation by known antiderivatives.

Exercise 3.9 For example, consider the function $f(x) = e^{x^2}$. Compute the integral

$$\int_0^1 f(x)\,dx$$

to 6 decimal places. For this purpose, determine the number n of necessary equidistant evaluation points by applying the above error estimate $(*)$ to each individual trapeze over the subintervals.

The Taylor Series

The Taylor series is probably the most important tool in numerical mathematics. It plays a major role in error estimation in approximation methods. For instance, the estimate (∗) above is based on the evaluation of a suitable Taylor series. In general, the Taylor series allows the approximation

$$f(x + h) \approx \sum_{i=0}^{n} \frac{h^i}{i!} f^{(i)}(x)$$

of a sufficiently often differentiable function f in a vicinity of x.

Here, we consider the exponential function. For this function we even have

$$e^x = \lim_{n \to \infty} \sum_{i=0}^{n} \frac{1}{i!} x^i, \quad x \in \mathbb{R}.$$

Exercise 3.10 Write a Python function

```
exp_approx(x)
```

for the approximation of e^x to 6 decimal places. Formulate the loop condition without reference to the true value. Try to use as few elementary arithmetic operations +, ∗, −, / as possible.

The Newton Method

The Newton method is a standard method for the numerical determination of zeros of nonlinear functions.

Let $f: \mathbb{R} \to \mathbb{R}$ be a continuously differentiable function. Consider the recursively defined sequence

$$x_{n+1} := x_n - \frac{f(x_n)}{f'(x_n)}, \quad x_0 \text{ given.}$$

If this sequence converges, then to a root of f. Again, this can be shown with the aforementioned Taylor-series expansion.

Exercise 3.11

(1) Write a Python function

```
newton(f, f_prime, x)
```

which applies the above Newton method to the initial value x_0 until $|x_{n+1} - x_n| < 10^{-7}$. To catch the case that the computation does not converge, it should be interrupted with an error message after 100 unsuccessful steps.

Note that in addition to the function f itself, the derivative f' must also be supplied explicitly.

(2) Test the procedure for $f(x) := x^2 - 2$ and the initial value $x = 1$. Compare the result to the exact solution $\sqrt{2}$.

Chapter 4
Python in Scientific Computation

The Python language itself contains only a limited number of mathematical functions. Fortunately, it is easy to seamlessly integrate comprehensive libraries. For instance, the computation of $\sin(\pi/2)$ could be performed as follows:

```
>>> from math import sin, pi
>>> sin(pi/2)  # Out: 1.0
```

The package `math` already includes a collection of important mathematical functions and constants, for example the standard trigonometric functions, the square root, the exponential, and different variants of the logarithm function. An advantage of `math` is that it is available in essentially every Python implementation. The precise content can be inquired with

```
>>> import math
>>> help(math)
```

However, the de facto standard for numerical mathematics is doubtlessly based on the combination of *NumPy* (for *Numerical Python*) and *SciPy* (for *Scientific Python*), together with *Matplotlib* for graphical visualization. Especially SciPy provides tools that extend Python to a programming environment comparable to the widespread commercial package MATLAB.

These approaches will be the subject of this chapter. This chapter is perhaps the single-most central chapter in the book. We return to the techniques and examples discussed here throughout.

The website `scipy.org` contains extensive documentation for all of the libraries. The books [2] and [5] can be recommended for further reading,

4.1 NumPy

There are several ways to include external libraries. One we have already seen above (and also encountered in the exercises in the last chapter). Another one is for instance (illustrated with `numpy` as example) to issue the command '`import numpy`'.

© Springer Nature Switzerland AG 2020
E. Smith, *Introduction to the Tools of Scientific Computing*, Texts in Computational Science and Engineering 25, https://doi.org/10.1007/978-3-030-60808-8_4

In that case, calls of NumPy components must be preceded by the prefix 'numpy.', such as numpy.sin(numpy.pi).

Often used is the form 'import numpy as np', which declares np to be an abbreviation for numpy, such that we can write np.sin(np.pi).

Here we often prefer the simplest form, namely 'from numpy import *', which implies that we can use the whole library content without prefix, and for instance write sin(pi).

As a first example, we write a NumPy program for the solution of quadratic equations. But before that we introduce the eminently important NumPy data type array.

NumPy Arrays

A NumPy array is similar to a Python list, however with two limitations. An array can only contain numbers, and the length of an array – once declared – cannot be changed. These apparent drawbacks however come along with a decisive advantage. The specialized arrays allow for significantly faster processing, a fact which certainly becomes relevant in data sets with, say, millions of elements. A further advantage is the algebraic structure of the arrays. They behave like mathematical vectors. They can be added, multiplied with scalars, or scalars may be added componentwise to all elements. As we will see, this idea of componentwise processing is a major feature.

After these preparations we come to our first example:

Example 4.1 (Quadratic Formula) We develop a program for the solution of quadratic equations $ax^2 + bx + c = 0$, $a \neq 0$, according to the formula

$$x_{1,2} = \frac{-b \pm \sqrt{b^2 - 4ac}}{2a}.$$

The following function solveq does this:

```
1 from numpy import sqrt, array
2 def solveq(a,b,c):
3     d = b**2 - 4*a*c
4     if d < 0: return 'No real-valued solutions'
5     w = sqrt(d)
6     return 1.0/(2*a)*(-b + array([w, -w]))
```

The root function sqrt in line 5 is not contained in the basic Python kernel, it belongs to NumPy. Note that it always returns only the positive root.

Lines 4 and 6 illustrate that the return value of a function is not limited to a single type, in line 4 it is a *string*, in line 6 a *pair of floats*, more precisely a NumPy array with two components.

We take a closer look at the expression in line 6. The root in line 5 is stored in ±-form in the list [w, -w] and then converted to a NumPy array by the function array(). Then the value -b is added component by component to each entry, which results in the array [-b+w, -b-w]. The computation is then terminated by multiplication of the array with the scalar '1.0/(2*a)'.

We test the program for $4x^2 - 12x - 40$:

```
7 print(solveq(4, -12, -40))   # Out: [ 5. -2.]
```

Exercise Modify solveq to an implementation of the roots method for the Parabola class in Sect. 3.9 in the last chapter.

Vectors and Matrices

Vectors and matrices are represented directly by NumPy arrays and can therefore be added to and multiplied by scalars.

In the following, we assume all required numpy components to be imported.

Vectors

In the example above we already indicated how NumPy arrays can be used to represent vectors. Here is a collection of basic vector operations:

```
1 >>> v = array([1, 2, 3])
2 >>> v[1]                    # access to components
3 >>> 2 * v                   # multiplication with scalar
4 >>> 3 + v                   # componentwise addition of scalar
5 >>> w = array([4, 5, 6])
6 >>> v + w                   # addition
7 >>> v * w                   # attention! componentwise multiplication
8 >>> inner(v,w)              # correct scalar product
9 >>> v @ w                   # correct scalar product, new in Python 3.5
```

In lines 1 and 5 the vectors v and w are created as Python lists and then converted to NumPy arrays by the numpy function array().

Lines 3, 4, 6 and 7 illustrate the component-based approach in NumPy arrays.

In line 8 the scalar product is implemented by a specific numpy function inner.

Note in line 9 that Python version 3.5 and higher provides '@' as an infix operator for matrix multiplication, and thus also for the scalar product.

Remark The two instructions print([1,2,3]) and print(array([1,2,3])) yield the same result [1,2,3]. However, the actual data type can be queried with the function 'type':

```
1 >>> type([1, 2, 3])         # Out: List
2 >>> type(array([1, 2, 3]))  # Out: numpy.ndarray
```

The response in line 2 is the internal name ndarray for the NumPy data type array. The prefix 'nd' stands for 'n-dimensional'.

Exercise What happens if we work directly with the *lists*, i.e. if we consider 2*[1,2,3] or [1,2,3]+[4,5,6] for lists?

Matrices

Matrices are two-dimensional arrays in the above sense and can also be represented directly as NumPy arrays:

```
1 A = array([[1, 2], [3, 4]])
2 A[0,1]                     # component
3 Atp = A.T                  # transpose of A
4 B = array([[5, 6], [7, 8]])
5 C = A + B                  # correct componentwise addition
6 D = A * B                  # attention! componentwise multiplication
7 E = dot(A,B)               # correct matrix multiplication
8 F = A @ B                  # correct matrix multiplication
```

The notation A.T in line 3 is an abbreviation for transpose(A).

In NumPy there is also a data structure matrix, specifically designed for matrices, such that, for instance:

```
>>> A = matrix([[1, 2], [3, 4]])
>>> B = matrix([[5, 6], [7, 8]])
>>> A * B    # here denotes correct matrix multiplication!
matrix([[19, 22],
        [43, 50]])
```

Other than that, the matrix type does not appear to offer any significant advantages. The prevailing opinion among users (especially in the expert discussions on the site stackoverflow.com) seems to be that one should stick to the more general array data type when dealing with matrices. In fact, with the new infix operator '@', there are even fewer reasons to use the dedicated matrix type.

Access to Matrix Components

As already seen for lists in general, it is often useful to extract parts from a given array. NumPy makes it possible to access not only individual elements of a matrix, but also arbitrary subblocks. For this, simply replace numbers with tuples when indexing.

We consider some useful NumPy methods specifically for matrices:

```
>>> A = array([[1, 2, 3, 4], [5, 6, 7, 8], [9, 10, 11, 12]]); A
array([[ 1,  2,  3,  4],
       [ 5,  6,  7,  8],
       [ 9, 10, 11, 12]])
>>> A[1,:]                   # select 2nd row (indices begin at 0)
array([5, 6, 7, 8])
>>> A[:,2]                   # select 3rd column
array([ 3,  7, 11])
>>> A[1:3,:]                 # two lines beginning with 2nd
array([[ 5,  6,  7,  8],
       [ 9, 10, 11, 12]])
>>> A[1:3, 1:3]             # submatrix
array([[ 6,  7],
       [10, 11]])
```

```
>>> A[:, (0, 2, 3)]
array([[ 1,  3,   4],
       [ 5,  7,   8],
       [ 9, 11,  12]])
>>> A[(0, 1),:]
array([[1, 2, 3, 4],
       [5, 6, 7, 8]])
>>> i = array([0, 2])
>>> A[:,i]
array([[ 1,   3],
       [ 5,   7],
       [ 9,  11]])
```

Flatten and Reshape

With the method `flatten` all elements of a matrix can be arranged linearly in an array, by default row by row from left to right:

```
1 >>> A = array([[1, 2, 3, 4], [5, 6, 7, 8], [9, 10, 11, 12]]);
2 >>> A.flatten()  # elements written in row-major order:
3 array([ 1, 2, 3, 4, 5, 6, 7, 8, 9, 10, 11, 12])
4 >>> A.flatten(order='F')  # in column-major order:
5 array([ 1,  5,  9,  2,  6, 10,  3,  7, 11,  4,  8, 12])
```

The parameter `'F'` in line 4 denotes that the matrix is read column by column and each column from top to bottom, as is usual in the Fortran programming language.

With `reshape(n,m)`, an array or matrix of compatible size can be reshaped into an $n \times m$ matrix with the same elements:

```
>>> a = array([1, 2, 3, 4, 5, 6])
>>> a.reshape(2, 3)  # row-major order:
array([[1, 2, 3],
       [4, 5, 6]])
>>> a.reshape(2, 3, order='F')  # column-major order:
array([[1, 3, 5],
       [2, 4, 6]])
```

Remark Note that the default option for `flatten` and `reshape` can also be specified explicitly as `order='C'`, referring to the representation order of matrices in the C programming language. The interested reader is referred to Sect. 6.4 in the C chapter.

The column-major option in `flatten` and `reshape` will be of particular importance in Sect. 4.8, where we solve partial differential Poisson equations with the so-called finite difference method. There we need to flatten a quadratic matrix U to a vector u with entries in column-major order. After some processing, the new u-values should be stored in U at the original positions.

The following example illustrates how the original positions can be regained:

```
>>> U = array([1, 2, 3], [4, 5, 6], [7, 8, 9]])
>>> u = U.flatten(order='F')
array([1, 4, 7, 2, 5, 8, 3, 6, 9])
```

```
>>> u.reshape(3, 3, order='F')
array([[1, 2, 3],
       [4, 5, 6],
       [7, 8, 9]])
```

Standard Matrices

Zero matrices, one matrices and identity matrices can be constructed by

```
>>> zeros((2,3))
array([[ 0.,  0.,  0.],
       [ 0.,  0.,  0.]])
>>> ones((2,3))
array([[ 1.,  1.,  1.],
       [ 1.,  1.,  1.]])
>>> eye(2)
array([[ 1.,  0.],
       [ 0.,  1.]])
```

Observe that 'eye' is pronounced like 'I'. That may be the reason for the choice of the name 'eye' for the identity matrix.

Advanced Construction Methods

Further methods for the definition of arrays and matrices are based on the already discussed *list comprehension*.

Example: Using the NumPy function arange(4) to define the array([0,1,2,3]) we get:

```
>>> array([i**2 for i in arange(4)])   # Out: array([0, 1, 4, 9])
```

As a somewhat larger example, a so-called *Toeplitz matrix*, i.e. a matrix in which every diagonal descending from left to right is constant can be constructed as follows:

```
>>> arc = lambda r, c: r - c
>>> array([[arc(r,c) for c in arange(4)] for r in arange(3)])
array([[ 0, -1, -2, -3],
       [ 1,  0, -1, -2],
       [ 2,  1,  0, -1]])
```

Remark The NumPy function arange is similar to the Python function range, with the difference that arange generates a NumPy array. The general form is arange(start,stop,step).

The Methods diag and kron, the Poisson Matrix

We show how to construct the so-called *Poisson* matrix, which we will use later to solve partial differential equations.

For a vector v (represented as a tuple, list, or array), diag(v) creates a square matrix with v as (main) diagonal. Similarly, diag(v,1) returns a matrix with v as first superdiagonal, and dually diag(v,-1) one with v as the first subdiagonal.

Example:

```
1  >>> n = 3
2  >>> v = 4*ones(n)
3  >>> diag(v)
4  array([[ 4.,  0.,  0.],
5         [ 0.,  4.,  0.],
6         [ 0.,  0.,  4.]])
7  >>> w = ones(n-1)
8  >>> diag(w, 1)
9  array([[ 0.,  1.,  0.],
10        [ 0.,  0.,  1.],
11        [ 0.,  0.,  0.]])
```

As a basic building block for our construction we need the tridiagonal matrix B in lines 14–16 below:

```
12  >>> sB = diag(w, 1) + diag(w, -1)
13  >>> B = diag(v) - sB; B
14  array([[ 4., -1.,  0.],
15         [-1.,  4., -1.],
16         [ 0., -1.,  4.]])
```

A very useful tool for generating block matrices is the *Kronecker product*. In general, it is defined as follows:

If A is an $m \times n$ matrix and B a $p \times r$ matrix, the Kronecker product $C = A \otimes B$ is the matrix

$$C = (a_{ij} \cdot B) = \begin{pmatrix} a_{11}B & \cdots & a_{1n}B \\ \vdots & \ddots & \vdots \\ a_{m1}B & \cdots & a_{mn}B \end{pmatrix},$$

for example:

$$\begin{pmatrix} 1 & 2 \\ 3 & 4 \end{pmatrix} \otimes \begin{pmatrix} 7 & 8 \\ 9 & 0 \end{pmatrix} = \begin{pmatrix} 1 \cdot \begin{pmatrix} 7 & 8 \\ 9 & 0 \end{pmatrix} & 2 \cdot \begin{pmatrix} 7 & 8 \\ 9 & 0 \end{pmatrix} \\ 3 \cdot \begin{pmatrix} 7 & 8 \\ 9 & 0 \end{pmatrix} & 4 \cdot \begin{pmatrix} 7 & 8 \\ 9 & 0 \end{pmatrix} \end{pmatrix} = \begin{pmatrix} 7 & 8 & 14 & 16 \\ 9 & 0 & 18 & 0 \\ 21 & 24 & 28 & 32 \\ 27 & 0 & 36 & 0 \end{pmatrix}.$$

In NumPy the corresponding operation is called kron.

With the matrices sB and B defined in lines 12 and 13, the Poisson matrix A is then constructed by

```
17  >>> I = eye(n)
18  >>> A = kron(I,B) + kron(sB, -I)   # Poisson matrix
```

For a given n the Poisson matrix has the shape (n^2, n^2).

4.2 Conjugate Gradient

As an example of how basic NumPy tools can be used successfully in mathematical programming, we discuss the method of *conjugate gradients* to solve linear equations. We will come back to the example several times in later chapters.

The method of conjugate gradients is a widely used iterative method for the solution of large linear equation systems $Ax = b$ for $n \times n$ matrices A, that are *symmetric* (i.e. $A^T = A$) and *positive definite* (i.e. $x \cdot Ax > 0$ for $0 \neq x \in \mathbb{R}^n$). A prominent example is the Poisson matrix constructed above.

The Quadratic Form

The method is based on the fact that for a symmetric and positive definite matrix A the solution x^* of $Ax = b$ coincides with the unique point x^*, at which the so-called *quadratic form*

$$f(x) := \frac{1}{2} x \cdot Ax - x \cdot b$$

assumes its minimal value.

It follows that the solution of the linear equation can be obtained by *gradient descent* along the function f.

The method of conjugate gradients is a particular variant of gradient descent, that after making an initial guess at the minimum x_0 of f, proceeds by iteratively choosing "search directions" characterized by suitable *mutually conjugate n*-dimensional vectors $p_k, k \geq 1$.

Hereby two vectors u, v are *conjugate* (with respect to A) if

$$u \cdot Av = 0.$$

Note that this actually defines a vector dot product, in respect to which u and v are then orthogonal.

The Method

For our construction we choose x_0 arbitrarily, and then set $p_1 := b - Ax_0$ as the residual vector necessary to reach b. Here we let $x_0 := 0$ and $p_1 := b$.

Now assume we have inductively determined p_k and x_k. Then we have to fix the new direction p_{k+1}. As indicated for p_1, a key to the choice is the *residual*

$$(1) \qquad\qquad r_k := b - Ax_k$$

that gives a measure of how close x_k is to the solution.

The method chooses p_{k+1}, such that it is *conjugate* to all prior p_i, i.e. such that

$$(2) \qquad p_{k+1} \cdot A p_i = 0, \quad i = 1, \ldots, k,$$

and moreover minimizes the distance

$$(3) \qquad \|p_{k+1} - r_k\|$$

in the usual Euclidean norm.

It can be shown that this is the case for

$$(4) \qquad p_{k+1} := r_k + \beta_k p_k \quad \text{with} \quad \beta_k := \frac{r_k \cdot r_k}{r_{k-1} \cdot r_{k-1}}.$$

Once the direction p_{k+1} is chosen, a new estimate x_{k+1} for the solution is obtained by minimizing f along the new direction. That is, x_{k+1} will be chosen by determining the scalar $\alpha_{k+1} = \alpha$ that minimizes

$$(5) \qquad \min_\alpha f(x_k + \alpha p_{k+1}).$$

With p_{k+1} in (4), it can be shown that the solution is then

$$(6) \qquad x_{k+1} := x_k + \alpha_{k+1} p_{k+1} \quad \text{for} \quad \alpha_{k+1} := \frac{r_k \cdot r_k}{p_{k+1} \cdot A p_{k+1}}.$$

In our solution program we will use that with the same scalar α_{k+1}, the residual in (1) can equivalently be expressed by the inductive definition

$$(7) \qquad r_{k+1} := r_k - \alpha_{k+1} A p_{k+1}.$$

Remark The method relies on a clever steepest descent, only once along every dimension, such that the exact result is reached in n steps, but a good approximate solution in practice much faster.

The construction resembles the Gram–Schmidt orthogonalization process, however with the advantage that the definition of each p_{k+1} depends only on the direct predecessor p_k.

A thorough but entertaining discussion can be found in the paper [11], obtainable from www.cs.cmu.edu.

Example 4.2 (Conjugate Gradient Method) First we import the required NumPy components and fix an example:

```
1 from numpy import array, zeros, sqrt
2 A = array([[  9.,    3.,   -6.,   12.], [  3.,   26.,   -7., -11.],
3            [ -6.,   -7.,    9.,    7.], [ 12.,  -11.,    7.,  65.]])
4 b = array([ 18.,   11.,    3.,   73.])
```

We set the initial values:

```
5 n = len(b)
6 x = zeros(n)
```

```
7  r = b.copy(); p = r.copy()              # copy to avoid side effects
8  rs_old = r @ r                          # needed for (4) and (6) above
```

Note that the numbers (n) in the comments refer to the equations above.

The iterative approximation is computed in the following loop:

```
9   for i in range(n):                     # n iterations to  exact result
10      Ap = A @ p                          # matrix-vector mult.
11      alpha = rs_old / (p @ Ap)           # needed for (6) and (7)
12      x += alpha*p                        # ... in (6)
13      r -= alpha*Ap                       # ... in (7)
14      rs_new = r @ r                      # update residual
15      if sqrt(rs_new) < 1e-10: break      # desired approximate precision
16      p = r + (rs_new / rs_old)*p         # used in (4)
17      rs_old = rs_new                     # prepare for next iteration
18  print(x)                               # Out: [ 1.   1.   1.   1.]
```

4.3 SciPy

SciPy (for Scientific Python) is a collection of mathematical functions and algorithms that build on the NumPy extension. With SciPy, Python becomes an interactive programming environment that is not inferior to commercial systems like MATLAB. Detailed explanations and tutorials can be found on the site `scipy.org`.

SciPy consists of various subpackages that cover different areas of scientific computing. In the following, we discuss the packages `linalg` for numerical linear algebra, `optimize` for the solution of non-linear equations as well as optimization and equilibrium problems, and `integrate` for integrals and ordinary differential equations. Especially the latter two packages profit much from graphical visualization-tools provided by the package `matplotlib`.

Throughout this chapter we assume the package `scipy` to be loaded in the form

```
from scipy import *
```

4.4 Linear Algebra

The SciPy package for processing linear algebra is strongly based on the NumPy data type `array`. Actually, SciPy can be conceived as a superset of NumPy, where NumPy provides the `array` data type and the most basic operations: indexing, sorting, reshaping, basic elementwise functions, etc. For historical reasons, NumPy also contains some core linear algebra functions, where the SciPy versions however tend to be more fully-featured. Note that all NumPy examples above will still work when 'numpy' is replaced by '`scipy`'.

In the following we assume the package `scipy.linalg` to be loaded in the form:

```
from scipy.linalg import *
```

Matrix Algorithms

Not surprisingly, the package `scipy.linalg` provides basic methods for matrix arithmetic.

Example:

```
A = array([[1, 2, 3], [1, 1, 1], [3, 3, 1]])
>>> det(A)   # determinant
2.0
>>> inv(A)   # matrix inverse
array([[-1.,   3.5, -0.5],
       [ 1.,  -4. ,  1. ],
       [ 0.,   1.5, -0.5]])
```

Linear Equations

Linear equations can easily be solved with SciPy:

```
1 >>> A = array([[1, 2, 3], [4, 5, 6], [7, 8, 8]])
2 >>> b = array([[2, 3, 5]])
3 >>> x = solve(A, transpose(b)); x
4 array([[-2.33333333],    # solution
5        [ 3.66666667],
6        [-1.        ]])
```

The verification confirms the solution:

```
7  >>> A @ x - transpose(b)
8  array([[  0.00000000e+00],
9         [  1.77635684e-15],
10        [  0.00000000e+00]])
```

Notice that we again used the infix operator '@' for matrix multiplication. We'll continue to do this in the following.

LU Decomposition

As known, an *LU decomposition* is a decomposition of a regular matrix A into the product of a lower triangular matrix L and an upper triangular matrix U. The row permutation generally required in this operation can be represented by a permutation matrix P.

We can compute an LU decomposition of A with the command `lu`, and store the components in a tuple `P, L, U` of matrices:

```
>>> A = array([[1, 2, 3], [1, 1, 1], [3, 3, 1]])
>>> P, L, U = lu(A)
>>> L  # lower triangular matrix with 1s in the main diagonal
array([[ 1.         ,  0.         ,  0.         ],
       [ 0.33333333,  1.         ,  0.         ],
       [ 0.33333333,  0.         ,  1.         ]])
>>> U
array([[ 3.         ,  3.         ,  1.         ],
       [ 0.         ,  1.         ,  2.66666667],
       [ 0.         ,  0.         ,  0.66666667]])
>>> P
array([[ 0., 1., 0.],
       [ 0., 0., 1.],
       [ 1., 0., 0.]])
```

We verify our decomposition with

```
>>> P @ L @ U
array([[ 1.,  2.,  3.],
       [ 1.,  1.,  1.],
       [ 3.,  3.,  1.]])
```

which in fact returns the original matrix A.

Cholesky Decomposition

The *Cholesky decomposition* factorizes a *symmetric positive definite* matrix A into an upper triangular matrix U and its transpose, such that $A = U^T U$.

For this purpose the package scipy.linalg offers the function cholesky. Example:

```
>>> A = array([[11, 12, 7], [12, 14, 10], [7, 10, 11]])
>>> U = cholesky(A); U
array([[ 3.31662479,  3.61813613,  2.11057941],
       [ 0.        ,  0.95346259,  2.47900273],
       [ 0.        ,  0.        ,  0.63245553]])
>>> U.T @ U  # returns A
```

Application: Method of Least Squares

The *method of least squares* is a method to solve overdetermined linear equation systems of the form

$$Ax = b, \quad A \in \mathbb{R}^{n \times m}, \ x \in \mathbb{R}^m, \ b \in \mathbb{R}^n, \ n > m.$$

The least squares method now determines a vector x such that the above equations are fulfilled "as close as possible". More precisely, x is calculated so that the *residual vector* $r := b - Ax$ becomes minimal with respect to the Euclidean norm $\| \cdot \|$.

We are therefore looking for a solution x to the problem

$$\text{Minimize} \quad \|Ax - b\| \quad \text{for } x \in \mathbb{R}^m,$$

instead of an exact solution.

Now, it is known that this vector x is the solution of the *Gaussian normal equation*

$$(*) \qquad\qquad\qquad\qquad A^T A x = A^T b.$$

The matrix $A^T A$ is symmetric and positive definite, hence satisfies the conditions of the Cholesky decomposition.

We show how Cholesky decomposition can be used in solving $(*)$ by computationally efficient forward and backward substitution with triangular matrices.

Example 4.3 We consider the overdetermined linear equation system

$$Ax = b \quad \text{with} \quad A = \begin{pmatrix} 3 & 6 \\ 4 & -8 \\ 0 & 1 \end{pmatrix}, \quad b = \begin{pmatrix} -1 \\ 7 \\ 2 \end{pmatrix},$$

and write a Python program to solve it using Cholesky decomposition:

```
>>> A = array([[3, -6], [4, -8], [0, 1]])
>>> b = array([-1, 7, 2])     # ".T" not necessary
>>> A.T @ A
array([[ 25, -50],
       [-50, 101]])
>>> U = cholesky(_); U
array([[ 5., -10.],          #  upper triangle
       [ 0.,   1.]])
>>> z = solve(U.T, A.T @ b)  #  forward substitution
>>> x = solve(U,z)           #  backward substitution
>>> x  # Out: array([ 5., 2.])
```

QR Decomposition

The QR decomposition factorizes a matrix A into a product $A = QR$, where Q is a matrix consisting of orthonormal column vectors, and R is an upper triangle matrix. Note that then $Q^T Q = I$, where I is the identity matrix of dimension corresponding to the column number of Q.

To compute QR decompositions, we use the function qr from `scipy.linalg`.

Here is an example for a square matrix A:

```
>>> A = array([[12, -51, 4], [6, 167, -68], [-4 , 24, -41]])
>>> Q, R = qr(A)
>>> Q
array([[-0.85714286,  0.39428571,  0.33142857],
       [-0.42857143, -0.90285714, -0.03428571],
       [ 0.28571429, -0.17142857,  0.94285714]])
```

```
>>> R
array([[-14.,   -21.,    14.],
       [  0.,  -175.,    70.],
       [  0.,     0.,   -35.]])
>>> Q.T @ Q   # returns I
>>> Q @ R     # returns A
```

QR Decomposition in Rectangular Matrices

QR decomposition is possible also when A is a general *rectangular* $n \times m$ matrix with, say, $n > m$.

Example 4.4 Let's see how far we get with SciPy. Consider once again the situation in Example 4.3:

```
>>> A = array([[3, -6], [4, -8], [0, 1]])
>>> Q, R = qr(A)
>>> Q
array([[-0.6,  0. ,  0.8],
       [-0.8,  0. , -0.6],
       [-0. , -1. ,  0. ]])
>>> R
array([[ -5.,   10.],
       [  0.,   -1.],
       [  0.,    0.]])
```

As we can see, qr has selected Q as a square 3×3 matrix and R in $(3, 2)$ shape by adding a 0 row at the bottom.

The Gaussian Normal Equation

We can also use QR decomposition to solve the Gaussian normal equation $(*)$. For this, however, we would prefer a factorization with a *regular square* matrix R, since then, observing $A = Q^T R$ and the identity $Q^T Q = I$, we can rewrite $(*)$ as

$$R^T R x = R^T Q^T b,$$

and, by taking inverses, as
$$x = R^{-1} Q^T b.$$

Actually, SciPy offers to generate the matrices Q and R in such a reduced ("economic") form, however with the warning that that operation mode will be deprecated in future releases. But presently we can still use

```
>>> A = array([[3, -6], [4, -8], [0, 1]])
>>> Q1, R1 = qr(A, mode='economic')
>>> b = array([-1, 7, 2])
>>> inv(R1) @ Q1.T @ b   # Out: array([ 5.,   2.])
```

to solve the Gaussian normal equation.

Once the above economic form is gone, however, we have to produce it from the standard form ourselves, like so:

```
>>> Q1 = Q[:,(0, 1)]; R1 = R[(0, 1),:]
```

Eigenvectors and Eigenvalues

An *eigenvector* of a square matrix A is a vector x that does not change its *direction* when multiplied with A, but is only *stretched* by a scalar factor λ, i.e. such that

(1) $$Ax = \lambda x.$$

The factor λ is a so-called *eigenvalue* of the matrix. A common procedure to compute eigenvectors and -values relies on the fact that all eigenvalues can be determined as roots of the *characteristic polynomial*

$$p(\lambda) := \det(A - \lambda I),$$

where I is the unity matrix with shape corresponding to A.

An eigenvector belonging to an eigenvalue is then obtained as a solution to the equation system (1).

For the calculation of eigenvalues and -vectors, the package `scipy.linalg` provides the function `eig`.

Example:

```
1 >>> A = array([[1, 2, 3], [4, 5, 6], [7, 8, 8]])
2 >>> e_val, e_vec = eig(A)
3 >>> e_val    # eigenvalues
4 array([ 15.55528261+0.j,  -1.41940876+0.j,  -0.13587385+0.j])
5 >>> e_vec    # eigenvectors
6 array([[-0.24043423, -0.67468642,  0.51853459],
7         [-0.54694322, -0.23391616, -0.78895962],
8         [-0.80190056,  0.70005819,  0.32964312]])
```

Note that eigenvalues are in general complex numbers, in our example, however, all are real, as one can recognize by the summand `0.j` where `j` denotes the imaginary unit.

We compute the product of A and the first eigenvector v_0, as well as the product with the associated eigenvalue:

```
 9 >>> (A @ e_vec)[:,0]
10 array([ -3.74002233,  -8.50785632, -12.47378976])
11 >>> e_val[0]*e_vec[:,0]
12 array([ -3.74002233,  -8.50785632, -12.47378976])
```

The matching results in lines 10 and 12 confirm the SciPy solution.

Sparse Matrices

In numerical mathematics a matrix is considered *sparse*, if there are so many entries of zeros that it is worth taking advantage of. Similarly, a vector consisting of mostly zeros is referred to as a *sparse vector*.

Later we will discuss partial differential equations and how to solve them in SciPy. In this context, significantly better computational performance is obtained if sparse instead of full matrices are used.

To deal with sparse matrices, SciPy provides the package sparse, which in the following is assumed to be imported in the form

```
>>> from scipy.sparse import *
```

To illustrate the idea we first produce a zero matrix of shape (1000, 1000) using the function lil_matrix:

```
>>> A = lil_matrix((1000,1000))  # class initialization of linked list
>>> A
<1000x1000 sparse matrix of type '<class 'numpy.float64'>'
        with 0 stored elements in LInked List format>
```

Technically, A is declared as a *linked list* (more about that in the chapter on the language C). Even if we have not yet stored any values in A, we can access the components. They however all return the value 0:

```
>>> A[111, 222]  # Out: 0.0
```

We illustrate the use of sparse matrices in the solution of large systems of linear equations. For this, we need to insert (sparsely many) values into A. To do so we use the NumPy function rand(n), which generates an array of n seemingly random numbers in the interval [0, 1).

```
1 >>> from numpy.random import rand
2 >>> A[0, :100] = rand(100)
3 >>> A[1, 100:200] = A[0, :100]
4 >>> A.setdiag(rand(1000))  # method in class lil_matrix
```

In line 2 an array of 100 random numbers is generated and then stored in the matrix A at the positions A[0,0] to A[0,99]. In line 3 these values are copied to the positions A[1,100] to A[1,199]. In line 4 the whole main diagonal is filled with random numbers (where, incidentally, the already existing value of A[0,0] is overwritten).

We then define a vector

```
>>> b = rand(1000)
```

for which we intend to solve the equation $Ax = b$.

To do this, we need an internal representation of A in a computationally more suitable format as "compressed sparse row" (see for instance the Wikipedia article Sparse_matrix). The conversion is achieved by

```
>>> A = A.tocsr(); A
<1000x1000 sparse matrix of type '<class 'numpy.float64'>'
        with 1199 stored elements in Compressed Sparse Row format>
```

The equation is then solved with the solution method `spsolve`, which is specially adapted to sparse matrices:

```
>>> from scipy.sparse.linalg import spsolve
>>> x = spsolve(A,b)
```

Eigenvalues

There are also specially adapted functions for calculating eigenvalues and vectors in sparse matrices. As an example we use `eigs` to determine the two smallest (with respect to absolute magnitude), as well as the three largest eigenvalues of the matrix A:

```
>>> from scipy.sparse.linalg import eigs
>>> e_val, e_vec = eigs(A, 2, which='SM')
>>> e_val_s  # Out: array([ 0.00041116+0.j,   0.00156261+0.j])
>>> e_val_l, e_vec_l = eigs(A, 3, which='LM')
>>> e_val_l  # Out: array([0.9997+0.j, 0.9964+0.j, 0.9961+0.j])
```

For details the reader may consult

```
>>> help(eigs)
```

The Methods diags and kron

We have already discussed the construction of Poisson matrices. Now, the Poisson matrix is typically applied when the basic size n is quite large, say $n = 50$. The Poisson matrix then has the shape (n^2, n^2). For $n = 50$ it is $(2500, 2500)$.

As is easy to see, the matrix is sparse. For reasons of storage (and, as we will see later, also for reasons of computation speed) it is therefore advantageous to consider a sparse implementation.

For this we need

```
>>> from scipy import *
>>> from scipy import sparse
>>> from scipy.sparse import diags, identity
```

The construction itself is now the same as before, only with adapted functions.

```
>>> n = 50
>>> v = 4*ones(n)
>>> w = ones(n-1)
```

The vectors v and w are defined exactly as before.

For the matrices we use the `sparse` variant of all constructors:

```
>>> sB = diags(w, 1) + diags(w, -1)
>>> B = diags(v) - sB
>>> I = identity(n)
>>> A = sparse.kron(I,B) + sparse.kron(sB, -I)
```

As a result we get:

```
>>> A  # Poisson matrix
<2500x2500 sparse matrix of type '<class 'numpy.float64'>'
        with 12300 stored elements in Compressed Sparse Row format>
```

Conversion Between Full and Sparse Matrices

If you want to inspect the contents of a sparse matrix, you use the method `toarray`, for instance

```
>>> B.toarray()
```

Conversely, you can create a sparse matrix from a full one *A* by simply specifying how it should be represented internally, for example:

```
>>> A = eye(3)
>>> Alil = sparse.lil_matrix(A)   # Linked List format
>>> Acsr = sparse.csr_matrix(A)   # Compressed Sparse Row format
```

4.5 Graphics with Matplotlib

Visualization of function graphs is an important part of scientific programming. The package *Matplotlib* contains comprehensive tools to that end. The official website `matplotlib.org` provides extensive documentation.

Example 4.5 As a simple example we sketch a coarse approximation to the sine function over the interval $[0, 2\pi]$.

First we import `sin` and `pi` from NumPy, as well as the function `linspace` to generate the grid:

```
1 from numpy import sin, pi, linspace
2 xvals = linspace(0, 2*pi, 9)
```

In line 2 the grid is generated, for which the function is to be "plotted" (i.e. represented graphically). In general, the NumPy function `linspace(a,b,n)` generates a list of *n* equidistant values of the form `[i*(b-a)/(n-1) for i in range(n)]`. In analogy to the counting in garden fences, *n* grid points determine $n - 1$ intervals.

If the last argument `n` is not specified, it is by default assumed as 50, such that `linspace(a,b)` is short for `linspace(a,b,50)`.

Actually, `linspace` does not generate a *list* of Python data type `list`, but rather a NumPy *array*. As previously explained, this enables a componentwise application of functions, in the present case `sin(xvals)` for `[sin(x) for x in xvals]`:

```
3 yvals = sin(xvals)
```

We then use the functions 'plot' and 'show' in the library `matplotlib.pyplot` to create and display the plot. Matplotlib contains various collections, including `pyplot`, the most common choice today:

```
4 from matplotlib.pyplot import plot, show
5 plot(xvals, yvals)
6 show()
```

The instruction in line 5 creates the plot. It interprets the first argument `xvals` as a list of *x*-values, and the second as the associated *y*-values. The individual points (x, y) are then connected by straight line segments to yield the desired graph.

In line 6 the function 'show' then displays the plot to the screen. This is illustrated in Fig. 4.1.

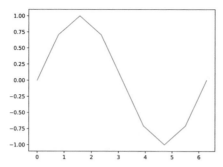

Fig. 4.1 The sine function represented on a grid of 9 equidistant supporting points.

Note that in many programming environments, such as e.g. Spyder, the plot is automatically displayed, the call to show hence not required.

The plot can also be saved to a file, say, graph.png like this:

```
from matplotlib.pyplot import savefig
savefig('graph.png')
```

Note however that the 'savefig' function cannot be called *after* 'show', since the latter consumes its argument, so to speak.

To obtain high-resolution graphics, the output format can be set to e.g. PDF, simply by changing the file extension to 'pdf' instead of 'png'. In fact, this is how the plots were prepared for printing in this book.

Example 4.6 We consider a somewhat larger example, where two functions are visualized within a common plot, the function $f(x) = x^2 e^{-x^2}$ and its derivative $f'(x) = 2x(1 - x^2)e^{-x^2}$ over the interval $[0, 3]$.
    The functions are defined by

```
1 def f(x): return x**2*exp(-x**2)
2 def dfdx(x): return 2*x*(1 - x**2)*exp(-x**2)
```

The interval to plot them over:

```
3 xvals = linspace(0, 3)  # recall: short for linspace(0, 3, 50)
```

To illustrate what is possible, we add some more details to the plot:

```
4 from matplotlib.pyplot import plot, xlabel, ylabel, legend, title
5 plot(xvals, f(xvals), 'r')
6 plot(xvals, dfdx(xvals), 'b--')
```

Here, 'r' and 'b' specify the color of the plots, '--' the dashed shape of the line.

```
7 xlabel('x-axis'); ylabel('y-axis')
8 legend(['f(x) = x^2 e^(-x^2)', 'd/dx f(x)'])
9 title('Function and derivative')
```

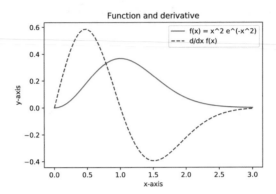

Fig. 4.2 Two functions plotted in a sufficiently fine grid

The result then looks like in Fig. 4.2. Note that the default number 50 for grid points in linspace usually yields an acceptably fine grid for a smooth representation.

Example 4.7 A function to be plotted may also be given by a number of value pairs of the form $(x, f(x))$. In this case, the x- and y-values must be distributed to two "parallel" arrays (or, as here, actually normal Python lists).

Example:

```
1 >>> pts = [(1, 1), (2, 4), (3, 9)]     # input
2 >>> xvals = [x for x, y in pts]        # x-components
3 >>> xvals  # Out: [1, 2, 3]
4 >>> yvals = [y for x, y in pts]        # y-components
5 >>> plot(xvals, yvals)                 # plot
6 >>> z = list(zip(xvals, yvals))        # reconstruction of the pts-list
7 >>> z  # Out: [(1, 1), (2, 4), (3, 9)]
```

In lines 2 and 4 the x- and y-components are collected in separate lists. Note that the pairs in pts are given as tuples, and therefore do not have to be enclosed in parentheses.

Line 6 shows how the components can be brought together again. The analogy to a *zipper* in clothing is mirrored in the function name zip. Note that zip works like the previously considered Python function range. The value pairs are generated one after the other. It is the function list() that combines them to a comprehensive whole.

4.6 Nonlinear Equations, Optimization

In the package optimize, SciPy offers a collection of functions for solving nonlinear equations and systems of such equations.

Single Equations

In a first example we consider the *bisection method* as well as *Newton's method*:

```
1 >>> from scipy.optimize import bisect, newton
2 >>> f = lambda x: x + 2*cos(x)
3 >>> bisect(f, -2, 2)   # Out: -1.0298665293221347
4 >>> newton(f, 2)       # Out: -1.0298665293222757
```

In line 2 we define the function, for which we want to determine the root. In line 3 we use the bisection method to find a root in the interval $(-2, 2)$. In line 4 we apply Newton's method with initial value 2.

Equation Systems

To compute solutions of equation *systems* we use the function fsolve. First, however, we apply fsolve to the same single equation as above:

```
>>> from scipy.optimize import fsolve
>>> fsolve(f, 0.3)  # initial value 0.3
array([-1.02986653])
```

Note that the output in array form already indicates that fsolve is intended for solving *systems* of multiple equations. Here is one:

$$y - x^3 - 2x^2 + 1 = 0$$
$$y + x^2 - 1 = 0.$$

Starting with an initial value $(1, 1)$, the system can be solved like this:

```
>>> def f(x): return [x[1] - x[0]**3 - 2*x[0]**2 + 1,
                       x[1] + x[0]**2 - 1]
>>> fsolve(f, [1, 1])  # Out: array([ 0.73205081,  0.46410162])
```

However, note that all solvers bisect, newton and fsolve determine the *root of a function* f. An *equation*, say, $x^2 = 2$, must be rewritten in the homogeneous form $f(x) = 0$ with $f(x) := x^2 - 2$.

Minimization

Often it is necessary to determine the minimum of a function. For this, we use the minimize operator in the package scipy.optimize.

As an example, we show how to use minimize to find the minimum of the so-called *Rosenbrock function*

$$f(x, y) := (1 - x)^2 + 100(y - x^2)^2.$$

We plot the function, here in a somewhat more elaborate style, simply to indicate what is possible:

```
from numpy import linspace, meshgrid
import matplotlib.pyplot as plt
from mpl_toolkits.mplot3d import Axes3D
from matplotlib import cm
fig = plt.figure()
ax = fig.gca(projection='3d')
x = linspace(-2, 2, 100)
y = linspace(-2, 4, 100)
x, y = meshgrid(x,y)
z = (1 - x)**2 + 100*(y - x*x)**2
surf = ax.plot_surface(x, y, z,
        rstride=1, cstride=1, cmap=cm.jet, linewidth=0)
plt.show()
```

The result is shown in Fig. 4.3.

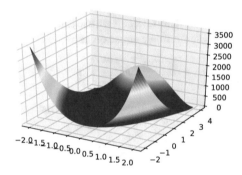

Fig. 4.3 The Rosenbrock function

The Rosenbrock function assumes its minimum value 0 at the point $(1, 1)$. The following program computes an approximate solution:

```
1 from scipy.optimize import minimize
2 def f(x): return (1 - x[0])**2 + 100*(x[1] - x[0]**2)**2
3 x0 = array([1.3, 0.7])
4 res = minimize(f, x0)
5 print(res.x)      # Out: [0.99999552 0.99999102]
6 print(f(rex.x))   # Out: 2.011505248124899e-11
```

In line 3 we choose an initial value. The solution 'res' returned in line 4 contains various entries, including the minimum coordinates in the array res.x.

To get a better control of the approximation, the function minimize can apply various approximation methods. In the present case we could for instance replace line 4 with

```
res = minimize(f, x0, method='nelder-mead', tol=1.e-8)
```

to obtain the exact result [1. 1.]. Use help(minimize) to see the available options.

4.7 Numerical Integration, Ordinary Differential Equations

In the package `integrate`, SciPy offers a large collection of tools for numerical integration and the solution of ordinary differential equations.

Integration

For the numerical computation of definite integrals, SciPy provides various functions, in particular the general-purpose tool 'quad'.

As a simple example, let's calculate the integral $\int_0^1 x^2 \, dx$:

```
>>> from scipy.integrate import quad
>>> f = lambda x: x**2
>>> ans, err = quad(f, 0, 1)   # outputs result and estimated error
>>> ans   # Out: 0.33333333333333337
>>> err   # Out: 3.700743415417189e-15
```

Also integrals from/to $\pm\infty$ can be calculated, for instance $\int_0^\infty e^{-x} \, dx$:

```
>>> f = lambda x: exp(-x)
>>> quad(f, 0, inf)   # Out: (1.000000, 5.842607e-11)
```

Note that we have rounded the result to fit it into one line.

Ordinary Differential Equations ODEs

The `integrate` package offers various functions for the numerical solution of systems of ordinary differential equations, including the solver `solve_ivp` for solving initial value problems IVPs and `solve_bvp` for boundary value problems BVPs.

Initial Value Problems, `solve_ivp`

As a simple example, we consider the *initial value problem*

$$(1) \qquad \begin{aligned} u'(x) &= 2x, \quad x \in [0, 1], \\ u(0) &= 0, \end{aligned}$$

with the exact solution $u_e(x) = x^2$.

We first illustrate the central idea in numerical approximation with the so-called *Euler method*.

In (1) it is obvious that a solution can be found by *direct integration*:

$$(2) \qquad u(x) = \int_{u(0)}^x u'(s) \, ds = \int_0^x 2s \, ds.$$

We fix a sequence $0 = x_0 < \cdots < x_{n-1} < x_n = 1$.

Then it is again obvious that the solution u satisfies the following equation system:

$$(3) \qquad u(x_{i+1}) = u(x_i) + \int_{x_i}^{x_{i+1}} 2s\,ds, \quad i = 0, 1, \ldots, n-1.$$

If we assume the evaluation points equidistant with distance $1/n$, and approximate the integral in (3) by $1/n \cdot 2x_i$, we obtain a discrete function u that assumes the values $u_i := u(x_i)$, inductively defined by

$$(4) \qquad u_0 = 0, \quad u_{i+1} = u_i + 2/n \cdot x_i, \quad i = 0, 1, \ldots, n-1.$$

This function u can now easily be computed by a Python program:

```
u = zeros(100)
n = 100
x = linspace(0, 1, n)
for i in range(n-1): u[i+1] = (u[i] + 2*x[i]/n)
from matplotlib.pyplot import plot
plot(x,u)
```

The Solver solve_ivp

Example 4.8 Equation (1) can also be solved directly with `solve_ivp`:

```
1  from scipy.integrate import solve_ivp
2  def dudx(x,u): return 2*x
3  epts = (0, 1)
4  u0 = [0]
5  sol = solve_ivp(dudx, epts, u0)
```

Line 2 specifies the form in which u' is passed to the solver. Line 3 sets the evaluation interval, specified by the tuple of endpoints. In line 4 the initial value is given in form of a *list*. This is necessary, since in systems of multiple equations, an initial value has to be specified for each.

The solution `sol` stores the evaluation points selected by the solver in `sol.t`, and the associated function values in `sol.y[0]`. In its basic form, the algorithm actually only returns 6 value pairs. We'll see below how to get finer solution sets.

More precisely, t is returned as an array of length 6, y as a 1×6 matrix, where the values for u are contained in y[0].

Remark The reason for the specific identifier t in `sol.t` reflects that initial value problems are often considered for systems that evolve along a *time* parameter.

We turn to a slightly less trivial example, where in particular we show how to specify finer evaluation meshes.

Example 4.9 We use `solve_ivp` to solve

$$u'(x) = x - u(x), \quad x \in [0, 5],$$
$$u(0) = 1.$$

Here is the program:

```
1 def dudx(x,u): return x - u
2 xvals = linspace(0, 5, 100)
3 epts = xvals[0], xvals[-1]
4 u0 = [1]
5 sol = solve_ivp(dudx, epts, u0, t_eval=xvals)
```

The solution is computed in line 5. The optional argument `t_eval=xvals` specifies the array of evaluation points used in the approximation. In line 3 we define the interval endpoints in terms of this array, namely by picking out the first and last element. Here we could of course also insert the bounds directly, for instance as a tuple `(0,5)`.

The solution can then be plotted:

```
6 plot(xvals, sol.y[0])  # note: xvals = sol.t
```

The result is illustrated in Fig. 4.4.

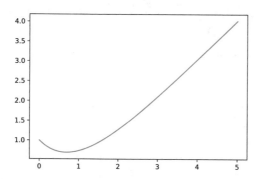

Fig. 4.4 Initial value problem solved with `solve_ivp`

Lotka–Volterra Equations

Systems of multiple first order differential equations can also be solved with the same solver `solve_ivp`.

As an example, we consider the Lotka–Volterra equations, a pair of first order equations, frequently used to describe the dynamics of predator–prey systems in biology.

A simplified model can be formulated as follows:

Example 4.10 Determine u and v such that

$$(*) \quad \begin{aligned} u'(t) &= u(t)\,(1 - v(t)), \\ v'(t) &= -v(t)\,(1 - u(t)), \quad t \in [0, 4\pi], \\ u(0) &= 1, \quad v(0) = 4. \end{aligned}$$

To solve the system with `solve_ivp`, we collect the unknown functions u and v in a pair $y = (u, v)$.

The equation system ($*$) then becomes

$$y' = (u \cdot (1 - v), \ -v \cdot (1 - u)), \quad y(0) = (1, 4).$$

Recalling that indices in Python begin at 0, we can now hand it over to the solver:

```
1 def dydt(t,y): return [y[0]*(1- y[1]), -y[1]*(1 - y[0])]
2 y0 = [1, 4]
3 tvals = linspace(0, 4*pi, 100)
4 epts = [0, 4*pi]
5 sol = solve_ivp(dydt, epts, y0, t_eval=tvals)
```

The array `sol.y` contains the solutions for u and v.

We plot both functions together:

```
6 from matplotlib.pyplot import plot
7 plot(tvals, sol.y[0])
8 plot(tvals, sol.y[1])
```

The result is shown in Fig. 4.5.

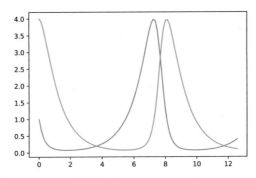

Fig. 4.5 Lotka–Volterra equations. The blue line represents u, the orange one v

Pendulum

Second order differential equations can be solved with `solve_ivp` by reduction to a *system* of *two* first order equations.

Example 4.11 As an example, we compute the angle θ of a pendulum that is subjected to gravity with friction.

The pendulum problem can be written as follows:

Determine θ such that

$$\theta''(t) + b\,\theta'(t) + c\sin(\theta(t)) = 0,$$

where b and c are positive constants, here assumed as $b = 0.25, c = 5.0$.

To solve the equation, we transform it into a system of two first-order equations. For this purpose we introduce the *angular velocity* $\omega(t) := \theta'(t)$ and obtain the equation system

$$(*) \qquad \begin{aligned} \theta'(t) &= \omega(t), \\ \omega'(t) &= -0.25\,\omega(t) - 5.0\sin(\theta(t)). \end{aligned}$$

We collect θ and ω in a function pair $y := (\theta, \omega)$. As initial conditions at $t = 0$ we assume the pendulum to be nearly vertical with $\theta(0) = \pi - 0.1$, and at rest, so that $\omega(0) = 0$. The pair of initial conditions is then given by $y_0 = (\pi - 0.1, 0)$.

In the following solution program, we first define a Python function `dydt(t,y)` that for $y = (\theta, \omega)$ returns the tuple $\frac{dy}{dt} = (\theta', \omega')$ of derivatives specified in $(*)$:

```
1  def dydt(t,y):
2      theta, omega = y
3      return omega, -0.25*omega - 5.0*sin(theta)
```

Then we declare the initial value and the solution interval:

```
4  y0 = (pi - 0.1, 0.0)
5  tvals = linspace(0, 10, 100); tspan = (0, 10)
```

The system can be now be solved:

```
6  sol = solve_ivp(dydt, tspan, y0, t_eval=tvals)
```

The return value `sol.y` is a 2×100 matrix. It contains the computed values of the unknown function θ in the first row, in the second the values of the "helper" function ω. The following plot shows them both:

```
7  from matplotlib.pyplot import plot, legend
8  plot(tvals, sol.y[0], label='theta(t)')
9  plot(tvals, sol.y[1], label='omega(t)')
10 legend(loc='best')
```

The result is illustrated in Fig. 4.6.

Boundary Value Problems, `solve_bvp`

Boundary value problems BVPs are in general harder to deal with than initial value problems, so also in Python. The solver `solve_bvp` was only added to the package `scipy.integrate` in 2016, so its use is not yet very widespread.

We briefly indicate how it works. As `solve_ivp` in the previous section, `solve_bvp` can only deal with first derivatives, so that we again have to translate a given second order problem into a system of two first order equations.

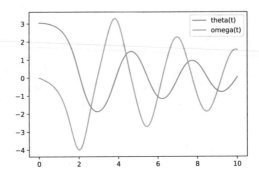

Fig. 4.6 Solution of pendulum equation

Example 4.12 Consider the BVP

$$u''(x) = 6x \text{ in } [0, 1],$$
$$u(0) = 0, \quad u(1) = 1$$

with the obvious solution $u(x) = x^3$.

To establish this in SciPy, we again introduce a helper function v with $u' = v$ and $v' = u''$. We represent the unknown functions by their derivatives $\frac{dy}{dx} = (u', v')$ as a Python tuple (y[1], 6*x), which declares that the derivative of y[0] is y[1] (corresponding to $u' = v$), and the derivative of y[1] is $6x$ (corresponding to $u'' = v' = 6x$).

This information is encoded in a Python function dydx, which the solver accesses repeatedly for different values during the approximation.

The program thus begins as follows:

```
1 def dydx(x,y): return (y[1], 6*x)
```

The boundary values $u(0) = 0$, $u(1) = 1$ are encoded in a function bc in *residual* form that describes what has to be subtracted to obtain a homogeneous equation system. In our case, the left boundary value is already 0, for the right one we have to subtract 1. Note that the arguments are again lists, where index position 0 refers to the original function u:

```
2 def bc(yl, yr): return (yl[0], yr[0]-1)
```

The interval under consideration, here $[0, 1]$, is initially encoded in the most basic possible from, as the pair of endpoints:

```
3 x_init = [0, 1]
```

In the matrix y_init, we can supply *initial-value guesses* for the unknown functions u and v at the points in the interval x_init, and pass them to the solver. Here we initialize all values to 0:

```
4 y_init = zeros((2, len(x_init)))
```

We can now apply the solver:

```
5 from scipy.integrate import solve_bvp
6 res = solve_bvp(dydx, bc, x_init, y_init)
```

The resulting approximation functions for both u and v are contained in `res.sol`.

For a vector of evaluation points, here

```
7 x = linspace(0, 1, 100)
```

the values for the function u can be retrieved by

```
8 u = res.sol(x)[0]
```

and then plotted with

```
9 from matplotlib.pyplot import plot
10 plot(x,u)
```

The result is shown in Fig. 4.7.

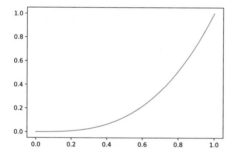

Fig. 4.7 BVP $u''(x) = 6x$, $u(0) = 0$, $u(1) = 1$ solved with `solve_bvp`

Remark For the curious reader: The meaning of the matrix y_init in lines 4 and 6 might still appear somewhat elusive. Actually, it serves a similar purpose as the initial value in Newton's method in the solution of nonlinear equations, namely to ensure efficient convergence, and moreover, in case of multiple solutions, to guide the solver to the desired one.

The row y_init[0] can assume initial function values $u(x)$ for the unknown function u, the row y_init[1] values for the helper function.

To make that point clear, we illustrate the best possible help we can give to the solver, namely to provide the exact solution itself. Replace line 3 by, say, 'x_init = linspace(0,1,5)', and assign the corresponding exact function values to the matrix y_init by inserting the following lines after line 4:

```
y_init[0] = x_init**3    # exact function value
y_init[1] = 3*x_init**2  # exact derivative value
```

Example 4.13 As a second simple example we consider the BVP

$$u''(x) = -2 \ \text{in} \ [0, 5],$$
$$u(0) = 0, \quad u(5) = 3.$$

The main difference from the previous example is the *constant* curvature -2. It must be recast to a *vector* of suitable length:

```
def dydx(x,y):
    z = -2*ones(len(x))
    return (y[1], z)
```

The rest of the problem specification is then essentially a repetition of the above:

```
def bc(yl, yr): return (yl[0], yr[0]-3)
x_init = [0, 5]
y_init = zeros((2, len(x_init)))
```

The system can now be solved as before.

Example 4.14 As a final example, we consider a "real world problem", the so-called *Bratu equation*

$$u''(x) = -e^{u(x)} \ \text{in} \ [0, 1],$$
$$u(0) = u(1) = 0,$$

which is known to have *two* solutions.

We begin as in the previous examples:

```
1 def dydx(x,y): return (y[1], -exp(y[0]))
2 def bc(yl, yr): return [yl[0], yr[0]]
```

In the present case we have to provide suitable hints to the solver, not least because we want to retrieve *both* solutions.

We first choose the argument interval as

```
3 x_init = linspace(0, 1, 3)
```

and introduce *two different matrices* y_init_1 and y_init_2 for the initial guesses:

```
4 y_init_1 = zeros((2, len(x_init)));
5 y_init_2 = zeros((2, len(x_init)))
6 y_init_2[0][1] = 3
```

In line 6 the initial guess for $u(0.5)$ is set to the value 3.

Correspondingly, we also compute *two* solutions based on the different initial guess values:

```
7 from scipy.integrate import solve_bvp
8 res_1 = solve_bvp(dydx, bc, x_init, y_init_1)
9 res_2 = solve_bvp(dydx, bc, x_init, y_init_2)
```

and represent both solutions in the same plot:

```
10  x = linspace(0, 1, 100)
11  u_1 = res_1.sol(x)[0]
12  u_2 = res_2.sol(x)[0]
13  from matplotlib.pyplot import plot, legend
14  plot(x, u_1, label='u_1')
15  plot(x, u_2, label='u_2')
16  legend()
```

The result is shown in Fig. 4.8.

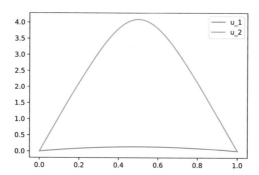

Fig. 4.8 Bratu equation with two solutions u_1, u_2

4.8 Partial Differential Equations

The solution of partial differential equations PDEs is one of the main concerns in numerical mathematics. By nature, the matter is significantly more complex than the IVPs and BVPs studied so far. There is no universal solver like `solve_ivp` or `solve_bvp`. SciPy provides useful support, but the main work has to be done "manually". In a later chapter we discuss the FEniCS project, which provides sophisticated methods for the automated solution of PDEs.

As a model example we discuss how *2D Poisson equations* can be solved in SciPy with the so-called *finite difference method*. We limit ourselves to *Dirichlet boundary conditions*, for a start even to *homogeneous* ones.

To this end, we consider the following PDE:

Determine a function $u: \Omega \to \mathbb{R}$, $\Omega := [0, 1] \times [0, 1]$, such that

$$
\begin{aligned}
-\Delta u &= f \quad \text{in } \Omega, \\
u &= g \equiv 0 \quad \text{on the boundary } \partial\Omega
\end{aligned}
$$
(1)

for a given function (often called *source function*) f, where we here assume the special case of a constant function $f \equiv 1$.

As usual, Δu is an abbreviation for the sum of the partial derivatives $\dfrac{\partial^2 u}{\partial x^2} + \dfrac{\partial^2 u}{\partial y^2}$.

For the construction of an approximate solution we use the general observation that for a twice continuously differentiable function φ and small h we have

$$\varphi''(x) \approx \frac{1}{h^2}\left(\varphi(x - h) - 2\varphi(x) + \varphi(x + h)\right).$$

For our function u in 2 arguments, we can add both approximations in the x- and y-directions to obtain

$$(2) \quad \Delta u(x, y) \approx \frac{1}{h^2}\left(u(x - h, y) + u(x, y - h) - 4u(x, y) + u(x + h, y) + u(x, y + h)\right).$$

Discretization

After these preparations we can now develop a discrete approximation for (1).

Let $h := 1/(n + 1)$ for a given n. We wish to determine the values for our desired function u at the points

$$u_{ij} = u(x_i, y_j), \quad \text{where} \quad x_i = ih, y_j = jh, \quad i, j = 0, \ldots, n + 1.$$

From (2) together with $-\Delta u = f \equiv 1$ in (1), we get the following system

$$(4) \quad \frac{1}{h^2}(-u_{i-1,j} - u_{i,j-1} + 4u_{ij} - u_{i+1,j} - u_{i,j+1}) = f(x_i, y_j) = 1, \quad i, j = 1, \ldots, n,$$

of n^2 equations as a discretization of the equation.

Each of the equations refers to a "5-point-stencil" around a u_{ij}, consisting of the adjacent points in the grid. Observing that $u_{ij} = 0$ for all boundary points with $\{i, j\} \cap \{0, n+1\} \neq \varnothing$, we remain with n^2 unknown u_{ij}. It follows that the system has a unique solution. (We will later deal with the general case of non-homogeneous boundary conditions.)

If we collect the unknown u_{ij} into a *column vector* u, in suitably ordered linear form, we can represent (4) as a matrix-equation

$$(5) \qquad\qquad\qquad \frac{1}{h^2} Au = b,$$

with an $n^2 \times n^2$ matrix A and a vector b of length n^2, containing the f-values in a corresponding order.

The matrix A obtains a particularly simple form, if the u_{ij} are ordered in column-major order, columnwise from left above to right below:

$$(6) \qquad\qquad u = (u_{1,1}, \ldots u_{n,1}, u_{1,2}, \ldots u_{n,2}, u_{1,3}, \ldots u_{n,n})^T.$$

In this case, each row v of A consists only of 0s, except for a value 4 on the main diagonal and maximally 4 entries with value -1, which are positioned precisely so that the scalar product $v \cdot u$ between that row and the unknown vector u amounts to the added-up values of the 5-point stencil around a u_{ij}. For instance, for $n = 3$, row 5 looks like this: $v = (0, -1, 0, -1, 4, -1, 0, -1, 0)$.

However, observe that, say, the first row, which refers to the stencil around $u_{1,1}$, consists of the entries $(4, -1, 0, -1, 0, \ldots, 0)$. For such border adjacent elements, there is no need to include the values that refer to the left and upper neighbor, since they are 0 by assumption.

Note that A actually turns out to be precisely the Poisson matrix from Sect. 4.1.

The definition of the vector b in this simple case with $f \equiv 1$ is trivial, i.e. we define b as the one-vector $(1, 1, \ldots, 1)^T$ of length n^2.

Solution Program

The equation system (5) can now be solved in SciPy. We first provide the grid:

```
1 n = 50                          # n x n inner grid points
2 h = 1/(n+1)                     # distance between grid points
```

Then we introduce the value matrix for the grid points, at first initialized with 0s:

```
3 u = zeros((n+2,n+2))          # value matrix for grid points 0-initialized
```

Next, we encode the source function f on the right-hand side in (1):

```
4 F = ones((n,n))               # encoding of right-hand side in (1)
```

The Poisson matrix A is constructed exactly as in Section 4.1:

```
5 v = 4*ones(n)
6 w = ones(n-1)
7 sB = diag(w, 1) + diag(w, -1)
8 B = diag(v) - sB
9 I = eye(n)
10 A = kron(I,B) + kron(sB, -I)
```

We come to the matrix equation (5). This requires a "flat" vector representation of the function values of f:

```
11 b = F.flatten(order='F')    # equation's right-hand side needs vector
```

The order='F' option ensures that the order in b corresponds to the column-major order of the u_{ij} according to (6). Actually, here it would not matter, since all f-values are equal by definition.

Then we can solve (5) with the operator 'solve' from scipy.linalg:

```
12 from scipy.linalg import solve
13 u_inner = solve(A, b*h*h)  # solution of lin. eq. syst. (5)
```

The solution u_inner has the form of a "flat" vector, and must be reshaped to a matrix, which then stores the inner values of the solution matrix u. More precisely, the values are stored in column-major order specified by (6), and must consequently be distributed accordingly. This is again achieved with the order='F' option:

```
14 u[1:n+1,1:n+1] = u_inner.reshape(n,n, order='F'))
```

The 0s on the boundary remain. As explained they represent the boundary condition in (1).

We are now ready to plot the solution:

```
15 lin = linspace(0, 1, n+2)
16 x, y = meshgrid(lin,lin)
17 from matplotlib.pyplot import figure, show
18 from matplotlib import cm
19 fig = figure()
20 from mpl_toolkits.mplot3d import Axes3D
21 ax = fig.gca(projection='3d')
22 surf = ax.plot_surface(x, y, u, rstride=1, cstride=1,
23                        cmap=cm.jet, linewidth=0)
24 show()
```

The result is shown in Fig. 4.9.

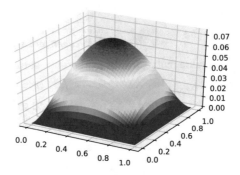

Fig. 4.9 Finite-difference approximation to Poisson equation (1)

Sparse Poisson Matrix

We show that a considerable gain in speed can be achieved in the above program by using sparse matrices.

To do this, we replace lines 7–10 in the program with

```
from scipy import sparse
from scipy.sparse import diags, identity
sB = diags(w, 1) + diags(w, -1)
B = diags(v) - sB
I = identity(n)
A = sparse.kron(I,B) + sparse.kron(sB, -I)
A = A.tocsr()
```

as well as lines 12 and 13 by

```
from scipy.sparse.linalg import *
u_inner = spsolve(A, b*h*h)
```

For example, the test computer used here for $n = 100$ takes approximately 0.034 seconds to compute the solution with a sparse Poisson matrix compared to 5.763 with a full matrix, i.e. a speed increase by a factor of about 165.

The time was measured by surrounding the lines calling the solvers as follows:

```
import time
start = time.time()
# The solver line goes here
end = time.time()
print(end - start)
```

Non-homogeneous Boundary Conditions

If the boundary function g in (1) above does not vanish everywhere, we speak of a *non-homogeneous* boundary condition.

We now discuss this general case, in which, as a further generalization, we no longer assume the source function f to be constant.

Consider once more the equation system (4) for a stencil around a grid point $(x_1, y_j) = (1h, jh)$, adjacent to a boundary point $p_{0j} = (0, y_j)$. The value u_{0j} is now given by $g_{0j} := g(x_0, y_j)$. The equation (4) then becomes

$$\frac{1}{h^2}(-g_{0j} - u_{1,j-1} + 4u_{1j} - u_{2j} - u_{2,j+1}) = f(0, y_j).$$

We can shift g_{0j} to the right-hand side, such that it now reads:

$$f_{0j} + \frac{1}{h^2}g_{0j}.$$

The remaining points adjacent to the boundary are treated similarly, in particular also the corner points that are adjacent to a horizontal *and* a vertical boundary, such as $(x_1, y_1) = (h, h)$.

This is all we need for the general solution.

The SciPy Program

This time we write a Python *function* that solves the PDE (1) and plots the solution for any given source function f and boundary-value function g on an $m \times m$ grid. The computation is based on sparse matrices:

```
1 def poisson_solver(f,g,m):
```

To make the `poisson_solver` self-contained, we import all needed components:

```
2      from scipy import ones, linspace, meshgrid
3      from scipy import sparse
4      from scipy.sparse import diags, identity, kron
5      from scipy.sparse.linalg import spsolve
```

The grid resolution is determined from the input m:

```
6      n = m-1; h = 1/(n+1)
7      lin = linspace(0, 1, n+2)
```

The solution matrix u is initialized with 0s:

```
8      u = zeros((n+2, n+2))
```

and then, evaluating the input function g, the boundary conditions are inserted:

```
9      for i in range(n+2):
10         u[i,0]   = g(lin[i], lin[0])
11         u[i,n+1] = g(lin[i], lin[n+1])
12     for j in range(1, n+1):
13         u[0,j]   = g(lin[0], lin[j])
14         u[n+1,j] = g(lin[n+1], lin[j])
```

The right-hand side of the equation system is initialized with 0s:

```
15     F = zeros((n,n))
```

and then populated with the source function f:

```
16     for i in range(n):
17         for j in range(n):
18             F[i,j] = f(lin[i+1], lin[j+1])
```

followed by the necessary modification of the boundary-adjacent points:

```
19     F[:,0]   += u[1:n+1, 0]   / h**2
20     F[:,n-1] += u[1:n+1, n+1] / h**2
21     F[0,:]   += u[0, 1:n+1]   / h**2
22     F[n-1,:] += u[n+1, 1:n+1] / h**2
```

Next comes the computation of the sparse Poisson matrix, exactly as before:

```
23     v = 4*ones(n)
24     w = ones(n-1)
25     sB = diags(w, 1) + diags(w, -1)
26     B = diags(v) - sB
27     I = identity(n)
28     A = kron(I,B) + kron(sB, -I)
29     A = A.tocsr()
```

We can then prepare for the solution process by reshaping F to a column vector, observing the column-major order required by (6):

```
30     b = F.flatten(order='F')
```

Then the actual solution is computed and stored in a vector u_inner:

```
31     u_inner = spsolve(A, b*h*h)
```

The flat solution vector is reshaped to a matrix, which is then inserted as a block of inner points into the solution matrix u:

```
32      u[1:n+1, 1:n+1] = u_inner.reshape(n,n, order='F')
```

Finally, the solution is ready to be plotted:

```
33      x, y = meshgrid(lin,lin)
34      import matplotlib.pyplot as plt
35      from matplotlib import cm
36      fig = plt.figure()
37      from mpl_toolkits.mplot3d import Axes3D
38      ax = fig.gca(projection='3d')
39      ax.plot_surface(x, y, u, rstride=1, cstride=1,
40                              cmap=cm.jet,linewidth=0)
41      plt.show()
```

Verification

We test the poisson_solver for the source function $f = 1.25\,e^{x+\frac{y}{2}}$ and the boundary-value function $g = e^{x+\frac{y}{2}}$:

```
from scipy import *
f = lambda x, y:  1.25*exp(x + y/2)
g = lambda x, y:  exp(x + y/2)
poisson_solver(f, g, 50)
```

and get the solution shown in Fig. 4.10.

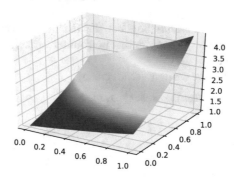

Fig. 4.10 Solution for non-homogeneous Poisson equation with source function $f = 1.25\,e^{x+\frac{y}{2}}$ and boundary-value function $g = e^{x+\frac{y}{2}}$

4.9 Round off: Random Numbers

We'll end our journey to numerical possibilities in Python with a quick visit to the casino. Many mathematical contexts (especially statistics) require numbers that ap-

pear to be the result of a random process. Previously, specially developed tables had
to be used for this purpose. This can easily be done today by programming. The
Python package random contains corresponding functions. We briefly show how random numbers can be used in numerical calculations. More precisely, we show how
to approximate the number π.

Monte Carlo Method for the Computation of π

At first glance, it is surprising that the number π can – so to speak – be computed by
tossing dice:

```
1 from random import random
2 samples = 1_000_000  # input,  '_' to increase readability
3 hits = 0
4 for _ in range(samples):
5     x, y = random(), random()
6     d = x*x + y*y
7     if d <= 1: hits += 1
```

Lines 5–7 are central. The random function generates a random real number in the
interval $[0, 1]$. In line 5 a random point p is chosen in the square $Q := [0, 1] \times [0, 1]$.

In line 6 it is checked, whether this p lies in the intersection S of the unit circle
and Q. If so, the counter 'hits' is incremented by 1.

Since the area ratio of $S : Q$ is exactly $\pi/4$, the ratio of hits to samples should
converge precisely to that value, under the assumption that the points are uniformly
distributed.

The loop counter in line 4 has the effect, that the loop is executed exactly 'samples'
many times. The symbol '_' denotes that no counter variable is used inside the loop,
hence it can be left out.

The result can then be written to the screen:

```
8 print(4*(hits/samples))  # Out: 3.143268
```

4.10 Exercises

Linear Equations

Exercise 4.1 The *Gauss-Seidel method* is an iterative method to solve a system of linear equations $Ax = b$. The matrix A is decomposed into a diagonal matrix D, a
strictly lower triangular component L, and a strictly upper triangular component U,
such that $A = L + D + U$.

Starting with an initial value $x^{(0)}$, the following sequence is then computed iteratively:
$$x^{(k+1)} := (D + L)^{-1}(b - Ux^{(k)}).$$

Consider the linear equation system

$$
\begin{pmatrix} 4 & 3 & 0 \\ 3 & 4 & -1 \\ 0 & -1 & 4 \end{pmatrix} \begin{pmatrix} x_1 \\ x_2 \\ x_3 \end{pmatrix} = \begin{pmatrix} 24 \\ 30 \\ -24 \end{pmatrix}.
$$

Perform three iteration steps using the Gauss-Seidel method, starting from the initial approximation $x^{(0)} = (3, 3, 3)^T$, and compare the result with the exact solution.

Nonlinear Equations

Exercise 4.2 Write

(1) an iterative Python function `bisect_it(f,a,b)`, and

(2) a recursive function `bisect_rec(f,a,b)`

to compute the root of a continuous function $f : [a, b] \to \mathbb{R}$, $f(a) < 0 < f(b)$, according to the *bisection method*.

Plot the function

$$
f(x) := \sin(4\,(x - 0.25)) + x + x^{20} - 1,
$$

and apply both `bisect_it` and `bisect_rec` to find a root.

Exercise 4.3 We continue with the Newton method in Exercise 3.11 in the last chapter. Extend the function `newton` to a Python function

```
newton_ext(f,x)
```

that now also computes the derivative values $f'(x)$.

Test both functions `newton` and `newton_ext` for

$$
f(x) := e^x + 2x,
$$

and compare the results.

Choose a suitable initial value x_0 by first plotting the graph of f.

Exercise 4.4 The equation $\cosh(x) = 2x$ has two solutions.

(1) Use the Newton method to determine both.

(2) Find an initial value x_0, for which the computation diverges.

Again it might be useful to inspect the graph.

Newton's Method in $\mathbb{R}^n$

For the computation of zeros of a function $f : \mathbb{R}^n \to \mathbb{R}^n$, $n > 1$, with Newton's method one has to "left-multiply" with the inverse of a Jacobian matrix instead of dividing by a derivative.

Exercise 4.5 Consider The function $f = (f_1, f_2, f_3) \colon \mathbb{R}^n \to \mathbb{R}^n$ with

$$f_1(x, y, z) := 9x^2 + 36y^2 + 4z^2 - 36,$$
$$f_2(x, y, z) := x^2 - 2y^2 - 20z,$$
$$f_3(x, y, z) := x^2 - y^2 + z^2.$$

(1) Define the Jacobian

$$J(v) := \left(\frac{\partial f_i(v)}{\partial v_j} \right)_{1 \le i, j \le 3}, \qquad v = (x, y, z).$$

(2) Solve the equation system $f(v) = 0$ approximately by considering the iteration sequence

$$u_{k+1} := u_k - J(u_k)^{-1} \cdot f(u_k)^T$$

with initial values $u_0 = (\pm 1.0, \pm 1.0, 0)$.

Integration

Exercise 4.6 Compute the integrals

$$\int_0^\pi \sin(x)\, dx, \qquad \int_0^2 3^{3x-1}\, dx, \qquad \int_0^1 e^{x^2}\, dx,$$

using different (at least three) methods in the package `scipy.integrate`.

Optimization

Exercise 4.7 Consider A and b in Example 4.2 for conjugate gradients in Sect. 4.2. Use the function `minimize` discussed in Sect. 4.6 to solve the equation $Ax = b$.

Ordinary Differential Equations, Initial Value Problems

Exercise 4.8 Solve the IVP specified by

$$u'(x) + u(x) = x,$$
$$u(0) = 1$$

with the method `solve_ivp` in the package `scipy.integrate`, and plot the solution together with the exact solution

$$u(x) = x - 1 + 2e^{-x}.$$

Exercise 4.9 Use solve_ivp to solve

$$(3x - 1)u'' - (3x + 2)u' - (6x - 8)u = 0,$$
$$u(0) = 2, \ u'(0) = 3.$$

Note that in the present case, an initial value is specified also for the *derivative u'*.

Ordinary Differential Equations, Boundary Value Problems

Exercise 4.10 Consider the boundary value problem

$$u''(x) = -|u(x)|, \quad x \in [0, 4],$$
$$u(0) = 0, \ u(4) = -2.$$

The equation has *two* solutions u_1, u_2. Use solve_bvp from scipy.integrate to determine both, and plot them. As a hint to the solver, which one to compute, you can use $u_1'(0) > 0, u_2'(0) < 0$.

Partial Differential Equations

Exercise 4.11 Consider the following program:

```
from numpy import *
n = 50
u = zeros((n,n))
x = y = linspace(0, 1, n)
u[0,:] = x*(1-x)
iter = 70
for _ in range(iter):
    for i in range(1, n-1):
        for j in range(1, n-1):
            v = u.copy()
            u[i,j] = (v[i+1,j] + v[i-1,j] + v[i,j+1] + v[i,j-1]) / 4
```

(1) Plot u[x,y] over the square determined by x and y.
(2) What is computed by the program? How and why does it work?

Chapter 5
Python in Computer Algebra

SymPy (for Symbolic Python) is a Python program package for the symbolic solution of mathematical problems. In the area of computer algebra, the commercial products Maple and Mathematica are probably the best known. However, SymPy does not have to hide from these professional environments. The expressiveness and performance is comparable.

SymPy is available for free. It has a programming community that encourages to participate and contribute to the development.

For introduction, the tutorial on the website `sympy.org` is recommended.

5.1 Symbolic Calculation, Numbers

Symbolic calculations are executed exactly, not approximately. Mathematical expressions that cannot be further processed, remain in symbolic form.

For example, compare the computation of $\sqrt{8}$ in SciPy and SymPy:

```
>>> import scipy, sympy
>>> scipy.sqrt(8)   # Out: 2.8284271247461903
>>> sympy.sqrt(8)   # Out: 2*sqrt(2)
```

`scipy.sqrt` calculates a numerical approximation, `sympy.sqrt` manipulates the expression, so that the exact root 2 of 4 is computed to $\sqrt{8} = \sqrt{4 \cdot 2} = 2\sqrt{2}$, but the remaining root of 2 is left as a symbolic expression.

In the following we assume that the SymPy package has been loaded in the form

```
>>> from sympy import *
```

To improve readability, we often also assume "pretty printing" is switched on with

```
>>> init_printing()
```

The effect of `init_printing` is that the output is printed in a mathematically formatted form, with the most suitable tools available on the machine. Here, we use the LaTeX formatting system.

E. Smith, *Introduction to the Tools of Scientific Computing*, Texts in Computational Science and Engineering 25, https://doi.org/10.1007/978-3-030-60808-8_5

Example:

```
>>> sqrt(8)
2√2
```

Number Types

To represent numbers, SymPy introduces three new number classes: `Integer`, `Float` and `Rational`. Some examples:

```
>>> a = Rational(1, 3); type(a)   # Out: sympy.core.numbers.Rational
>>> b = 9*a; type(b)              # Out: sympy.core.numbers.Integer
```

A numerical approximation of an exact symbolic expression can, for instance, be requested like this:

```
>>> sqrt(8).evalf(50)
2.8284271247461900976033774484193961571393437507539
>>> type(_)  # Out: sympy.core.numbers.Float
```

The method `evalf` returns an output of SymPy type `Float`. The number of decimals can be chosen as desired. `evalf()` without argument sets the output precision to 15 digits (Why?).

Symbolic Expressions

The main feature of computer algebra systems is that expressions built from symbolic variables can be manipulated.

In SymPy it is necessary to express such variables explicitly as *symbols*, e.g:

```
>>> x = Symbol('x')
```

The variable x now contains the SymPy symbol 'x' as value:

```
>>> x         # Out x
>>> type(x)   # Out: sympy.core.symbol.Symbol
```

In many other computer algebra systems – such as Maple or Mathematica – this declaration is not necessary. There, a variable to which no value has been assigned, is automatically interpreted as a symbol. This is not possible in Python, however, since variables can only be declared together with value assignments.

Remark The convention that a variable-name should coincide with the represented symbol, is meaningful but not compelling. It would also be possible to assign

```
>>> abc = Symbol('xyz')
```

A sequence of symbols can be declared with the function `symbols`:

```
>>> x, y, z = symbols('x y z')   # note lower-case initial letter s
```

Another equivalent option is to give the instruction

```
>>> from sympy.abc import x, y, z
```

In fact, with 'from sympy.abc import *' all Latin letters and Greek letters such as alpha, beta etc. are initialized as SymPy symbols.

Convention From now on, when needed we always assume variables x, y, z to be declared as symbols in one of the ways above.

Symbols are interpreted as number variables, so that simple algebraic rules are applied automatically:

```
>>> x + y + x - y   # Out: 2*x
```

sympify

To create a SymPy expression we can also convert a Python string, say

```
1 >>> str_expr = 'x**2 + 3*x - 1/2'
2 >>> type(str_expr)  # Out: str
```

With the command sympify we then get

```
3 >>> expr = sympify(str_expr)
4 >>> expr          # Out: x**2 + 3*x - 1/2
5 >>> type(expr)   # Out: sympy.core.add.Add
```

The use of sympify is not required very often, since basic arithmetic expressions containing only symbol variables are automatically understood as SymPy terms. However, some caution is required with *fractions*. We illustrate what happens for the following fractions:

```
>>> f1 = 1/3           # type: float
>>> f2 = sympify(1/3)  #      sympy.core.numbers.Float
>>> f3 = sympify(1)/3  #      sympy.core.numbers.Rational
>>> f4 = sympify('1/3') #     sympy.core.numbers.Rational
```

To ensure exact computations in expressions with division, it must be ensured that the result is of type Rational, since already *one* float or Float resulting from a division spreads through the entire computation. To be on the safe side, it is sufficient that one of the arguments in each division is of a well-behaved type.

Actually, the easiest way to achieve this, is to explicitly declare one of the involved integers as Rational, for instance, Rational(1)/3 or 1/Rational(3).

Value Assignments

Value assignments to symbolic variables are made with the method subs, for instance

```
>>> ((x + y)**2).subs(x,1)  # Out: (y + 1)**2
```

A symbol can also be replaced by another one, for example, y for x:

```
>>> ((x + y)**2).subs(x,y)   # Out: 4*y**2
```

Note that obvious algebraic simplifications have again been made automatically.

The substitution of more complex terms is also possible:

```
>>> ((x + y)**2).subs(x, y + z)   # Out: (2*y + z)**2
```

And also multiple substitutions:

```
>>> expr = x**3 + 4*x*y - z
>>> expr.subs([(x, 2), (y, 4), (z, 0)])   # Out: 40
```

Advanced Transformations

The real power of computer algebra systems, however, lies in the ability to mechanically transform even more complex expressions.

Here is an example, in which SymPy does not automatically recognize any obvious transformation. The expression is returned unchanged:

```
>>> (x + x*y)/x   # Out: (x*y + x)/x
```

simplify

In such cases often the help of the most general SymPy simplification method simplify already suffices. In the example above we get:

```
>>> simplify((x + x*y)/x)   # Out: y + 1
```

The effect of simplify is to try out a large number of algebraic transformation rules, and then to use the supposedly most meaningful one, according to a heuristic method. In general, however, there will be no canonical normal form nor an obvious goal of manipulation. In addition, it is a generally *undecidable* problem whether two algebraic expressions can be transformed into one another at all.

trigsimp

In many cases, a transformation can be achieved by more specific methods, for example in expressions involving trigonometric functions:

```
>>> trigsimp(sin(x)/cos(x))   # Out: tan(x)
```

expand

Often useful is the 'expand' method to generate a representation as a sum of multiplicative terms.

In polynomials, this yields a canonical form as a sum of monomials:

```
>>> expand((x + y)**3)   # Out: x**3 + 3*x**2*y + 3*x*y**2 + y**3
```

In many cases, user-provided information can be helpful:

```
>>> expand(cos(x + y), trig=True)
-sin(x)*sin(y) + cos(x)*cos(y)
```

Without the hint `trig=True`, the expression `cos(x + y)` is returned unchanged.

factor, collect, ...

For polynomials in particular, there are various other useful transformations.

The method `factor` factorizes a polynomial into irreducible factors over the rational numbers, for instance:

```
>>> factor(x**3 - x**2 + x - 1)   # Out: (x - 1)*(x**2 + 1)
```

The method `collect` collects and orders an expression according to term powers, here for the powers of x:

```
>>> expr = x*y + x - 3 + 2*x**2 - z*x**2 + x**3
>>> collect(expr, x)   # Out: x**3 + x**2*(-z + 2) + x*(y + 1) - 3
```

Functions

Symbolic expressions can be used to define functions. However, some care must be taken. The reader is invited to explain what happens in the following example:

```
>>> expr = x**2
>>> def f(x): return x**2
>>> def g(x): return expr
>>> def h(x_var): return expr.subs(x, x_var)
>>> f(1), g(1), h(1)   # Out: (1, x**2, 1)
```

Function definitions can also contain symbolic parameters, for instance:

Given the expression `expr = a*x**2`, we can define the function $f(x) := ax^2$, retaining the symbolic parameter a:

```
1 >>> a = Symbol('a')
2 >>> expr = a*x**2
3 >>> def f(varx): return expr.subs(x, varx)
4 >>> f(2)   # Out: 4*a
```

Often we don't need the actual conversion to a function. For instance, for plotting, SymPy provides a `plot` function that refers directly to term expressions.

Continuing with the example above, we can plot the graph for $a = 2$ like this:

```
5 >>> plot(expr.subs(a, 2), (x, 0, 1))
```

The result is shown in Fig. 5.1.

Actually, 'plot' returns the plot as an instance of the SymPy `Plot` class, which can then be stored in a variable for later use.

With the `save()` method the plot can also be saved to an output file:

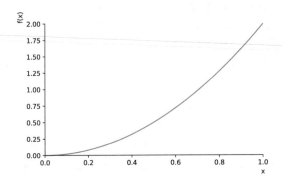

Fig. 5.1 Graph generated by the SymPy `plot` function

```
6 >>> graph = plot(expr.subs(a, 2), (x, 0, 1))
7 >>> graph.save('graph.pdf')
```

lambdify

If, however, we do need a "normal" function representation, for instance because we want to use Matplotlib for plotting, we can use the SymPy command `lambdify` to perform the conversion.

Example 5.1 We plot the function $f(x) := x^2 \sin(x)$ over the interval $[0, \pi]$ with the plot functions of both SymPy and Matplotlib.

First the SymPy term:

```
1 import sympy as sym
2 x = sym.Symbol('x')
3 expr = x**2*sym.sin(x)
```

plotted in SymPy:

```
4 sym.plot(expr, (x, 0, sym.pi))
```

Then the term is 'lambdified' to a Python function:

```
5 f = sym.lambdify(x, expr)
```

and plotted with Matplotlib:

```
6 import numpy as np
7 x = np.linspace(0, np.pi)
8 import matplotlib.pyplot as plt
9 plt.plot(x, f(x))
```

Fig. 5.2 shows the SymPy plot generated in line 4 to the left, and the one drawn by Matplotlib in line 9 to the right.

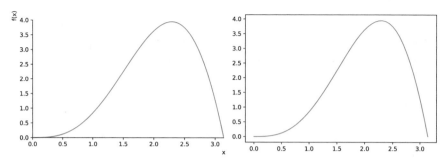

Fig. 5.2 Plots generated by SymPy (left) and Matplotlib (right)

5.2 Equation Systems

The simplest (and conventional) way to solve an equation in SymPy uses the operator solve. For example, the equation $x^2 = 1$ is solved by

```
>>> solve(Eq(x**2, 1), x)    # Out: [-1, 1]
```

Here Eq(.,.) denotes the equation, and x is the variable, for which the equation is to be solved. Note that unlike other computer algebra systems, the notations 'x**2 = 1' or 'x**2 == 1' do not work, because '=' and '==' are already reserved for other purposes in Python.

The equation above can equivalently also be written as $x^2 - 1 = 0$, which can then be solved by

```
>>> solve(x**2 - 1, x)
```

If the first argument consists of only one term, solve assumes that it is to be tested for equality to 0.

An even simpler way to write this, is

```
>>> solve(x**2 - 1)
```

since here the unknown x is clear from the context.

The search range for a solution can be specified, e.g. limited to real-valued solutions:

```
>>> solve(x**4 - 1, x)
[-1, 1, -I, I]    # I is the SymPy symbol for the imaginary unit
>>> x = Symbol('x', real=True)
>>> solve(x**4 - 1, x)    # Out: [-1, 1]
```

Equation *systems* can also be solved:

```
>>> solve([Eq(x + 5*y, 2), Eq(-3*x + 6*y, 15)], [x,y])
{y: 1, x: -3}
```

Remark Note that the result is returned in form of a dictionary of type dict, a Python data type briefly discussed in Sect. 3.5 in the Python chapter.

solveset and `linsolve`

It is often recommended to use the function `solveset` for solving general equations and the function `linsolve` specifically for linear equations.

`solveset`

For example, the equation $x^2 = x$ has the solutions 0 and 1, both found by `solve` as well as `solveset`:

```
>>> solve(x**2 - x)      # Out: [0, 1]
>>> solveset(x**2 - x)   # Out: {0, 1}
```

But note the difference: `solve` provides a *list* of the single solutions, `solveset` the entire solution *set* {0, 1}. The difference becomes clearer, when we consider the equation $x = x$. The solution of this equation is the largest possible number set, which for SymPy is the whole set of complex numbers $\mathbb{C}$.

```
>>> solve(x - x)      # Out: []
>>> solveset(x - x)   # Out: S.Complexes
```

The `solve` method fails, because the solution cannot be represented as a finite list, whereas the method `solveset` correctly returns `S.Complexes`, in pretty print $\mathbb{C}$.

With additional hints, `solveset` can also provide more precise characterizations of the solution sets, such as

```
1 >>> solveset(x - x, x, domain=S.Reals)  # Out: Reals
2 >>> solveset(sin(x) - 1, x, domain=S.Reals)
3 ImageSet(Lambda(_n, 2*_n*pi + pi/2), Integers())
```

The solution in line 3 becomes clearer, if we assume "pretty printing" to be switched on by the command 'init_printing()'. Then, instead of the cryptic expression, we get the understandable formulation

$$\left\{ 2n\pi + \frac{\pi}{2} \mid n \in \mathbb{Z} \right\}$$

Remark For readers who are familiar with the LaTeX formatting system, we note that for printing in this book the pretty-print outputs of SymPy were prepared with the SymPy function `latex`, the last one for example with

```
>>> print(latex(solveset(sin(x) - 1, x, domain=S.Reals)))
\left\{2 n \pi + \frac{\pi}{2}\; |\; n \in \mathbb{Z}\right\}
```

`linsolve`

Here is just a simple example. We return to the solution of linear equations in the context of matrix equations.

```
>>> linsolve([x+y+z-1, x+y+2*z-3], (x,y,z))  # Out: {(-y-1, y, 2)}
```

The solution set consists of all tuples (x, y, z) of the form $x = -y - 1$, $z = 2$, with arbitrary values for y.

Remark The curly braces suggest that the output is of type 'set', as discussed in the basic Python chapter. In fact, it is, but of a particular SymPy type 'sets.FiniteSet'.

5.3 Linear Algebra

SymPy is familiar with the usual methods of linear algebra. Both symbolic and numerical calculations are supported.

Matrices and Vectors

In SymPy, matrices are generated as lists of row vectors:

```
>>> M = Matrix([[1, -1], [3, 4], [0, 2]])
```

Elements of a matrix M are accessed by M[i,j], where indices again begin at 0.

A flat list of *single* elements is interpreted as a *column* vector:

```
>>> v = Matrix([1, 2, 3])
```

The components are accessed by v[0] etc.

The standard matrix operations +, * and ** apply:

```
>>> M = Matrix([[1, 2, 3], [3, 2, 1]])
>>> N = Matrix([0, 1, 1])
>>> M*N
Matrix([
[5],
[3]])
```

The transpose of a matrix M is generated by transpose(M) or M.T, the inverse of a square matrix M by M**-1:

```
>>> Matrix([[1, 2,], [3, 4]])**-1
Matrix([
[ -2,    1],
[3/2, -1/2]])
```

The determinant is computed by M.det(). The shape (i.e. the number of rows and columns) of a matrix can be queried with M.shape.

Single rows and columns are accessed by row and col, the first row of M for instance by M.row(0).

SymPy provides a special mechanism for generating vectors or matrices whose entries can be described by a function.

As an example, consider the famous Hilbert matrix:

```
>>> def f(i,j): return 1/(1+i+j)
>>> Matrix(3, 3, f)
Matrix([
[  1, 1/2, 1/3],
[1/2, 1/3, 1/4],
[1/3, 1/4, 1/5]])
```

Special matrices can be created, as in SciPy, by eye, zeros, etc.

Vector Space Bases

SymPy linear algebra tools are conveniently used to establish linear dependency or independence between vectors. We discuss some typical examples.

Example 5.2 We show that the following vectors in $\mathbb{R}^3$ form an orthonormal basis:

$$v_1 := \frac{1}{\sqrt{3}}(1,1,1), \quad v_2 := \frac{1}{\sqrt{2}}(1,0,-1), \quad v_3 := \frac{1}{\sqrt{6}}(1,-2,1).$$

In SymPy representation:

```
1 >>> v1 = Matrix([1, 1, 1])  / sqrt(3)
2 >>> v2 = Matrix([1, 0, -1]) / sqrt(2)
3 >>> v3 = Matrix([1, -2, 1]) / sqrt(6)
```

For the scalar product calculated with SymPy we get

$$\langle v_i, v_j \rangle = \begin{cases} 1, \text{ if } i = j, \\ 0, \text{ else.} \end{cases}$$

For instance:

```
4 >>> v1.dot(v1)  # Out: 1
5 >>> v1.dot(v2)  # Out: 0
```

Now, if $\{v_1, v_2, v_3\}$ is an orthonormal basis, then for every vector $x = (a, b, c) \in \mathbb{R}^3$, taking the sum of the projections we have

$$x = \sum_{i=1}^{3} \langle x, v_i \rangle v_i.$$

Let's try to verify this in SymPy:

```
6 >>> a, b, c = symbols('a b c')
7 >>> x = Matrix([a,b,c])
8 >>> y = v1.dot(x)*v1 + v2.dot(x)*v2 + v3.dot(x)*v3
```

Here is the result in pretty printing:

```
9 >>> y
```

$$\left[\begin{array}{c} \frac{\sqrt{2}}{2}\left(\frac{\sqrt{2}a}{2} - \frac{\sqrt{2}c}{2}\right) + \frac{\sqrt{3}}{3}\left(\frac{\sqrt{3}a}{3} + \frac{\sqrt{3}b}{3} + \frac{\sqrt{3}c}{3}\right) + \frac{\sqrt{6}}{6}\left(\frac{\sqrt{6}a}{6} - \frac{\sqrt{6}b}{3} + \frac{\sqrt{6}c}{6}\right) \\ \frac{\sqrt{3}}{3}\left(\frac{\sqrt{3}a}{3} + \frac{\sqrt{3}b}{3} + \frac{\sqrt{3}c}{3}\right) - \frac{\sqrt{6}}{3}\left(\frac{\sqrt{6}a}{6} - \frac{\sqrt{6}b}{3} + \frac{\sqrt{6}c}{6}\right) \\ -\frac{\sqrt{2}}{2}\left(\frac{\sqrt{2}a}{2} - \frac{\sqrt{2}c}{2}\right) + \frac{\sqrt{3}}{3}\left(\frac{\sqrt{3}a}{3} + \frac{\sqrt{3}b}{3} + \frac{\sqrt{3}c}{3}\right) + \frac{\sqrt{6}}{6}\left(\frac{\sqrt{6}a}{6} - \frac{\sqrt{6}b}{3} + \frac{\sqrt{6}c}{6}\right) \end{array} \right]$$

This is not yet very convincing, but we get the desired confirmation by applying

```
10 >>> simplify(_)
```

$$\left[\begin{array}{c} a \\ b \\ c \end{array} \right]$$

Example 5.3 We show that the polynomials

$$p_1(x) :\equiv 1, \quad p_2(x) := x - 1/2, \quad p_3(x) := x^2 - x + 1/6$$

constitute a basis of the vector space $\mathbb{P}_2$ of polynomials of degree 2. To prove this, for an arbitrary such polynomial $q(x) = ax^2 + bx + c$ we have to find coefficients c_1, c_2, c_3 such that

$$q(x) = \sum_{i=1}^{3} c_i p_i(x).$$

In SymPy we proceed as follows:

```
1 >>> from sympy import symbols, sympify, solve
2 >>> x, a, b, c, c1, c2, c3 = symbols('x a b c c1, c2, c3')
3 >>> p1 = 1
4 >>> p2 = x - Rational(1)/2  # Rational to ensure correct division
5 >>> p3 = x**2 - x + Rational(1)/6
6 >>> q = a*x**2 + b*x + c
7 >>> expr = q - (c1*p1 + c2*p2 + c3*p3)
8 >>> solve(expr, [c1, c2, c3])
```

Lines 3–5 define the basis polynomials. In line 6 we fix an arbitrary polynomial of degree 2 that is to be represented as a linear combination of the basis functions. In line 7 the *difference polynomial* is generated. In line 8 the values of c_1, c_2, c_3 are determined, such that the coefficients of all x-powers in the difference polynomial evaluate to 0.

SymPy's answer in pretty print:

$$\{c1 : c + \frac{1}{2}b + \frac{1}{3}a, \; c2 : b + a, \; c3 : a\}$$

It shows that the coefficients of the basis functions can, in fact, be selected to generate a representation of $q(x) = ax^2 + bx + c$ as a linear combination of the p_i, as desired.

Note again that the result is returned in form of a Python dictionary of type `dict`.

Example 5.4 We show how to check if the three vectors

$$u_1 = (1, 0, 2), \quad u_2 = (0, 1, 1), \quad u_3 = (1, 2, -1)$$

are linearly independent. To this end we consider the matrix

```
A = Matrix([[1, 0, 2], [0, 1, 1], [1, 2, -1]])
```

We can verify that A is regular by checking that `A.det()` yields a non-zero result -5. Furthermore, `A.columnspace()` shows that the space of the column vectors has the maximal dimension 3. The kernel `A.nullspace()` of the linear mapping induced by A is empty, which is still another way to establish regularity.

We determine the linear combination of the u_i to represent an arbitrary vector, say

```
>>> b = Matrix([8, 2, -4])
```

For this purpose, we use one of the customary solvers for matrix equations:

```
>>> u = A.LUsolve(b)
```

We use the result $u = \frac{1}{5}(8, -6, 16)$ to verify the computation:

```
>> Eq(A*u, b)   # Out: True
```

Note that in difference to normal float comparisons, we are here safe, since all computations are exact.

Example 5.5 Finally, an example of a matrix with non-empty kernel:

```
>>> C = Matrix([[1, 3, -1, 2], [0, 1, 4, 2],
                [2, 7, 2, 6], [1, 4, 3, 4]])
>>> C.nullspace()
```

$$\left[\begin{bmatrix} 13 \\ -4 \\ 1 \\ 0 \end{bmatrix}, \begin{bmatrix} 4 \\ -2 \\ 0 \\ 1 \end{bmatrix} \right]$$

Note that SymPy chooses the basis vectors to consist only of integer entries.

Eigenvectors and Eigenvalues

Recall the definition of eigenvectors and eigenvalues in Sect. 4.4 in the last chapter. To compute these, SymPy provides the methods `eigenvects` and `eigenvals`. Example:

```
1 >>> M = Matrix([[3, -2, 4, -2], [5, 3, -3, -2],
2                 [5, -2, 2, -2], [5, -2, -3, 3]])
3 >>> M.eigenvals()   # Out:  {3: 1, -2: 1, 5: 2}
```

The outcome of line 3 is a dictionary with the individual eigenvalues 3, -2 and 5 as keys, and their multiplicities as values.

This result is confirmed by inspecting the *characteristic polynomial* of M in factorized representation:

```
4 >>> lamda = Symbol('lamda')     # spelling to avoid conflict
5 >>> factor(M.charpoly(lamda))   # Output in pretty-print:
```

$$(\lambda - 5)^2(\lambda - 3)(\lambda + 2)$$

A representation of the eigenvalues together with the corresponding eigenvectors is obtained by

```
6 >>> M.eigenvects()
```

$$\left[\left(-2, \; 1, \; \left[\begin{bmatrix} 0 \\ 1 \\ 1 \\ 1 \end{bmatrix} \right] \right), \; \left(3, \; 1, \; \left[\begin{bmatrix} 1 \\ 1 \\ 1 \\ 1 \end{bmatrix} \right] \right), \; \left(5, \; 2, \; \left[\begin{bmatrix} 1 \\ 1 \\ 1 \\ 0 \end{bmatrix}, \begin{bmatrix} 0 \\ -1 \\ 0 \\ 1 \end{bmatrix} \right] \right) \right]$$

The components of this nested list can be accessed using component notation, for example the tuple in position 0 of the enclosing list like this:

```
 7 >>> comp0 = M.eigenvects()[0]
 8 >>> val0 = comp0[0]
 9 >>> vec0 = comp0[2][0]
10 >>> Eq(M*vec0, val0*vec0)  # Out: True
```

The last line checks whether the eigenvalue property is actually fulfilled. Note again that we can trust the verdict here, since computations in SymPy are exact.

$n \times n$ Matrices with $n \geq 5$

The problem of computing the eigenvalues of an $n \times n$ matrix corresponds to the problem of finding the roots of a polynomial of degree n. One of the most famous results of mathematics in the nineteenth century is that it is impossible to find a closed formula for solving general equations of degree ≥ 5, which Niels Henrik Abel proved in 1824. Therefore the attempt to determine exact eigenvalues of a large matrix with SymPy will not be successful.

To illustrate this, let's consider the 5×5 variant of the Hilbert matrix already mentioned above and try to determine the eigenvalues with the SymPy command eigen-vals:

```
def f(i,j): return 1/(1+i+j)
H5 = Matrix(5, 5, f)
H5.eigenvals()
```

SymPy aborts with an error message, i.e. the attempt to find a symbolic solution has failed.

5.4 Calculus

With its basic on-board equipment, SymPy can already handle limit values, as well as differential and integral computations.

Limits

SymPy knows a number of rules for the calculation of limit values, such as for example $\lim_{x=0} \sin(x)/x$:

```
>>> limit(sin(x)/x, x, 0)   # Out: 1
```

or $\lim_{x=\infty} 1/x$:

```
>>> limit(1/x, x, oo)       # Out: 0
```

Here 'oo' is the SymPy symbol for ∞.

One-sided limits can be calculated by

```
>>> limit(1/x, x, 0, '+')   # Out: oo
>>> limit(1/x, x, 0, '-')   # Out: -oo
```

Differential Calculus

`diff(f(x),x)` calculates the derivative of a function $f(x)$. SymPy knows the derivatives of all standard functions, as well as the usual derivation rules.

Example:

```
1 >>> diff(sin(2*x), x)   # Out: 2*cos(2*x)
2 >>> diff(tan(x), x)     # Out: tan(x)**2 + 1
3 >>> h = Symbol('h')
4 >>> limit((tan(x+h) - tan(x))/h, h, 0)   # Out: tan(x)**2 + 1
```

In line 4, for comparison, the derivative $\tan'$ is computed as

$$\tan'(x) = \lim_{h=0} \frac{\tan(x+h) - \tan(x)}{h}$$

The `diff` operator can also be written as suffix in the dot notation:

```
>>> expr = sin(2*x)
>>> expr.diff(x)   # Out: 2*cos(2*x)
```

Partial derivations are also possible, here for example $\dfrac{\partial^2 f(x, y)}{\partial x \partial y}$ for $f(x, y) = x^4 y$:

```
>>> diff(x**4*y, x, y)   # Out: 4*x**3
```

The nth derivative of a function can be characterized as follows:

```
>>> n = 3
>>> diff(sin(2*x), x, n)   # Out: -8*cos(2*x)
```

Integration

SymPy knows the integrals of the standard elementary functions, and can deal with the usual integration rules.

Example:

```
>>> integrate(log(x), x)          # indefinite integral
x*log(x) - x
>>> integrate(x**3, (x, -1, 1))    # definite integral
0
```

Also more complex integrals, such as

$$\int_{-\infty}^{\infty} \int_{-\infty}^{\infty} e^{-x^2-y^2} \, dx \, dy$$

can be evaluated symbolically:

```
>>> integrate(exp(-x**2-y**2), (x, -oo, oo), (y, -oo, oo))  # Out: pi
```

As is easily guessed, pi is the SymPy symbol for π. With init_printing the output would in fact be precisely the glyph π.

Remark Note, however, that the SymPy symbol for Euler's number is E and not e:

```
>>> exp(1)  # Out: E
```

Series Expansion

series(f(x),x,x0,n) calculates the coefficients of the terms $(x - x_0)^i$, $i < n$, in the Taylor expansion of the function $f(x)$ around the point x_0.

Example:

```
>>> series(cos(x), x, pi, 4)  # Output in pretty-print:
```

$$-1 + \frac{(x - \pi)^2}{2} + O\left((x - \pi)^4 ; x \to \pi\right)$$

The Landau symbol O here indicates that powers ≥ 4 are not evaluated.

For the special case $x_0 = 0$ and $n = 6$ we can also use 'series(f(x), x)':

```
>>> series(exp(x), x)
```

$$1 + x + \frac{x^2}{2} + \frac{x^3}{6} + \frac{x^4}{24} + \frac{x^5}{120} + O\left(x^6\right)$$

Like the diff operator before, the series operator can also be also written in suffix notation:

```
>>> exp(x).series(x)
```

5.5 Ordinary Differential Equations

The operator `dsolve` is used to symbolically solve ordinary differential equations. As a simple example, consider the boundary value problem $u''(x) = 6x$ in the interval $[0, 1]$ with boundary conditions $u(0) = 0,\ u(1) = 1$, discussed in Example 4.12 in the last chapter.

For the representation in SymPy we first need a variable-symbol 'u' for the unknown function u. We get it through

```
1   >>> u = Function('u')
```

or equivalently u = `symbols('u', cls=Function)`.

The differential equation is expressed by

```
2 >>> ode = Eq(u(x).diff(x,x), 6*x)
```

The general solution is then immediately obtained:

```
3 >>> gen_sol = dsolve(ode, u(x))
4 >>> gen_sol   # Out: Eq(u(x), C1 + C2*x + x**3)
```

Presently, SymPy does not yet deal satisfactorily with boundary conditions, so we do it manually, as follows:

```
5 >>> C1, C2 = symbols('C1 C2')
6 >>> expr = gen_sol.rhs   # right-hand side
7 >>> b = solve([Eq(expr.subs(x,0),0), Eq(expr.subs(x,1),1)], [C1, C2])
8 >>> sol = expr.subs(b)
9 >>> sol   # Out: x**3
```

The solution can then be plotted with SymPy's `plot` command:

```
10 plot(sol, (x, 0, 1))
```

The graph will have the same form as the one seen in the last chapter.

As a somewhat less trivial example, we consider the equation

$$u''(x) - 2u'(x) + u(x) = \sin(x),$$

which can be solved by

```
>>> u = Function('u')
>>> ode = Eq(u(x).diff(x,x) - 2*u(x).diff(x) + u(x), sin(x))
>>> dsolve(ode, u(x))
```

Assuming "pretty printing", we get the result as

$$u(x) = (C_1 + C_2 x)\, e^x + \frac{1}{2}\cos(x)$$

which can then be further processed by hand.

If there is no solution in closed form (or SymPy does not find it), `dsolve` returns a symbolic "intermediate result".

Example: Consider the equation

$$u'(x)(1 - \sin(u(x))) = 0 :$$

```
>>> u = Function('u')
>>> ode = Eq(u(x).diff(x)*(1 - sin(u(x))), 0)
>>> dsolve(ode, u(x))  # Out: Eq(u(x) + cos(u(x)), C1)
```

5.6 Galerkin Method

The Galerkin method is the basis for a powerful approach in the automated solution of differential equations, that we will discuss extensively in a later chapter, the FEniCS project.

The idea is to approximate the solution u of a differential equation by a linear combination

$$u \approx \sum_{i=1}^{n} c_i \varphi_i \quad \text{of given basis functions } \varphi_1, \dots, \varphi_n.$$

To illustrate how this method can be implemented in SymPy, we discuss a simple example.

The Model Equation

We consider a 2-point boundary value problem with *(homogeneous) Dirichlet boundary conditions*

(1)
$$-u'' = f \quad \text{over } [0, 1], \quad \text{where } f(x) := x\,(x + 3)\,e^x,$$
$$u(0) = u(1) = 0.$$

The Variational Problem

The Galerkin method relies on the so-called *variational form* of a differential equation. We briefly explain the idea.

Let V be the set of integrable functions $v \colon [0, 1] \to \mathbb{R}$ with $v(0) = v(1) = 0$.

If $-u'' = f$, it is then trivially clear that for every *test function* $v \in V$:

(2)
$$-\int_0^1 u'' v = \int_0^1 f v.$$

Now the crucial observation is that also the converse is true: if (2) holds for every such test function, then $-u'' = f$.

This is best seen by contraposition, as in the following exercise:

Exercise Assume $g, h \in C[0,1]$, $g \neq h$. Show that there is a $v \in C[0,1]$, $v(0) = v(1) = 0$, such that $\int_0^1 gv \neq \int_0^1 hv$.

These observations yield an equivalent formulation for (1) as a *variational problem*:

Determine u with $u(0) = u(1) = 0$ such that (2) holds for every $v \in V$.

It is common to refer to the left-hand side in (2) with $a(u, v)$, whereby it should be noted that a is bilinear in the arguments u and v, due to the linearity of the derivative and integral. Likewise, the right-hand side, which does not depend on u, is linear in v. It is often referred to as $L(v)$.

The Method

Consider a trial function $u = \sum_{i=1}^{4} c_i \varphi_i$ built from the the basis functions

$$\varphi_1(x) := x(1 - x),$$
$$\varphi_2(x) := x(1/2 - x)(1 - x),$$
$$\varphi_3(x) := x(1/3 - x)(2/3 - x)(1 - x),$$
$$\varphi_4(x) := x(1/4 - x)(1/2 - x)(3/4 - x)(1 - x).$$

Note that by inserting $u = \sum_{i=1}^{4} c_i \varphi_i$ in (2) and recalling the definition of the forms a and L above, for each test function v we get

(3)
$$\sum_{i=1}^{4} c_i a(\varphi_i, v) = a(u, v) = L(v).$$

We choose a suitable *finite* set of test functions v_j to generate a system of equations to determine the c_i. More precisely, we choose the basis functions φ_j themselves. From (3) we then obtain the linear equation system

(4)
$$\sum_{i=1}^{4} c_i a(\varphi_i, \varphi_j) = L(\varphi_j), \quad j = 1, \dots, 4,$$

of four equations to determine four unknown c_i.

The SymPy Program

The equation system (4) can now be solved with SymPy. We begin with the import of the SymPy components:

```
1  from sympy import *
```

Then we define the basis functions:

```
2 x = Symbol('x')
3 phi0 = x*(1 - x)
4 phi1 = sympify('x*(1/2 - x)*(1 - x)')
5 phi2 = sympify('x*(1/3 - x)*(2/3 - x)*(1 - x)')
6 phi3 = sympify('x*(1/4 - x)*(1/2 - x)*(1/4 - x)*(1 - x)')
7 phi = [phi0, phi1, phi2, phi3]
```

Again, note the use of `sympify`, which causes all fractions to be of the SymPy type `Rational`.

We define a SymPy function $a(u, v)$ to compute the integrals on the left-hand side in (3):

```
8 a = lambda u,v: -integrate(diff(u,x,2)*v, (x, 0, 1))
```

All these values are then collected in the so-called *stiffness matrix* A:

```
9 A = Matrix(4, 4, lambda i, j: a(phi[i], phi[j]))
```

Correspondingly, we define the SymPy function $L(v)$ to compute the right-hand side in (3):

```
10 f = x*(x + 3)*exp(x)
11 L = lambda v: integrate(f*v, (x, 0, 1))
```

Similarly to the stiffness matrix, we can then collect the values in a vector (called *load vector*) b.

```
12 b = Matrix([L(phi[j]) for j in range(4)])
```

Recall that a matrix generated from a list of single elements is interpreted as a *column* vector.

This is all we need to determine the coefficients c_i from the matrix equation $Ac = b$:

```
13 c = A.LUsolve(b)
```

With the coefficients c_i we define the function $u := \sum_{i=1}^{4} c_i \varphi_i(x)$ to represent the approximate solution:

```
14 u = sum(c[i]*phi[i] for i in range(4))
```

Note that here the standard Python function `sum`, briefly mentioned in the basic Python chapter, can also be used to add up SymPy terms.

The solution is then plotted with the SymPy function `plot`:

```
15 plot(u, (x, 0, 1))
```

The result is shown in Fig. 5.3.

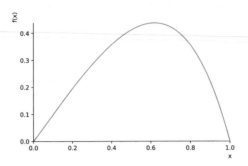

Fig. 5.3 Galerkin approximation for boundary value problem (1)

Exact Solution

To see how close our approximation is, for comparison we calculate the exact solution by direct integration:

```
16 from sympy import symbols, solve
17 C1, C2 = symbols('C1, C2')
18 temp = integrate(-f) + C1        # here var x unique from context
19 expr = integrate(temp, x) + C2   # ... here x needed
20 sol = solve([expr.subs(x, 0), expr.subs(x, 1)], [C1, C2])
21 u_e = expr.subs([(C1, sol[C1]), (C2, sol[C2])])
22 plot(u_e, (x, 0, 1))
```

In fact, the deviation between approximate and exact solution does not show up in the plot. In such cases, it is often useful to plot the *difference*:

```
23 plot(u - u_e, (x, 0, 1))
```

The result is shown in Fig. 5.4.

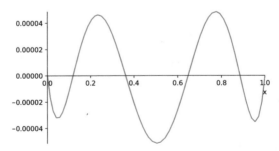

Fig. 5.4 Difference between exact solution of BVP (1) and Galerkin approximation

Chapter 6
The C Language

C is one of the most commonly used programming languages. In fact, we have already made extensive use of a C program. The standard implementation of the Python interpreter is written in C.

C was developed in the early 1970s by Dennis Ritchie in the United States, as the successor to a language B, for the development of the operating system Unix.

The main application of C is in system programming. Language constructs are designed to correspond closely to machine instructions. This allows for extremely efficient programming techniques, but with the disadvantage that the language often appears cryptic to beginners.

C was originally not intended for numerical calculation. There were special languages for this, such as Fortran (for *Formula Trans*lation). In the meantime, however, an increasing number of extensive Fortran libraries for scientific and numerical computation have been ported to C and its extensions such as C++.

In contrast to Python, C programs are *compiled*. The development is as follows: The program is created with an editor and saved in a file, usually with the extension '.c'. The compiler translates it into machine language and includes the required additional components from the program libraries. The result is an independent program that can be run directly on the machine.

Integrated Development Environment

For program development on personal computers, the Eclipse *integrated development environment* (IDE), which was originally written for the Java programming language, also seems to be suitable for C. Another popular environment is Microsoft's Visual Studio. Apple's Xcode is particularly recommended for macOS. All examples in this chapter have been tested in Xcode for the C variant C99.

© Springer Nature Switzerland AG 2020
E. Smith, *Introduction to the Tools of Scientific Computing*, Texts in Computational Science and Engineering 25, https://doi.org/10.1007/978-3-030-60808-8_6

Command Line Compiler

It is also possible to compile and run C programs from the command line. As an example we consider Ubuntu Linux, which we will use often in later chapters. When a program is stored in a file, e.g. B. `test.c` in the home directory of the user, it can be translated into an executable program with the Terminal command

```
$ gcc test.c
```

Note that `gcc` is one of the standard compilers used for C.

The program can then be run with

```
$ .\a.out
```

Here `a.out` is the default name automatically assigned to the executable program, which is again stored in the users home directory.

The prefix '`.\`' is an abbreviated representation of the access path to the home directory.

If another file name, say, `testprog` is desired, it can be specified with the output option '`-o`':

```
$ gcc test.c -o testprog
$ .\testprog
```

6.1 Basics

Example 6.1 As a first example, let us look again at the 'greeting 'Hello World'':

```
1 #include <stdio.h>          // contains the printf function
2 int main() {                // every C-prog. begins with main
3    printf("Hello World\n");
4    return 0; }              // not required
```

In line 3 the string enclosed in quotation marks is output to the screen. As in Python, the control character '\n' causes a line break. The function `printf` (for "print format-ted") is not part of the core language, but rather provided by the library `stdio` (for "standard input output") in line 1. The `#include` statement is equivalent to the state-ment '`from ... import *`' in Python.

Every C program starts at the line with the `main` function, which must therefore occur exactly once in the program. Formally, `main` is a function that returns a value of type `int`, here the value 0 in line 4. The 0 stands for 'properly terminated'.

In contrast to Python, an instruction block (here lines 3 and 4) is enclosed in curly braces. Indentation is only used to improve readability. In the following we adhere to the Python style, not emphasizing the curly braces. Also, if allowed by C, we will not enclose single-line statements in braces.

Instructions in C are always terminated by a semicolon.

Comments up to the end of a line are indicated by '`//`', comment blocks by inclu-sion in a pair '`/*, */`'.

Numbers and Arithmetic Operations

Basic number types and arithmetic operations are similar to what we've seen in Python.

Number Representations

The default data types for numbers are int for integers and float for floating-point numbers. On most systems, int has a word length of 32 bits = 4 bytes, float one of 32 or 64 bits = 8 bytes.

There are also different size variants, for integers for instance char, typically 1 byte in size, and 'unsigned long', typically 8 bytes in size. For floating-point numbers, in addition to float, double with a length of 8 bytes is common.

We will simply speak of int and float also referring to the other variants.

Arithmetic Operators

C knows the basic arithmetic operators +, -, *, /, as well as '%' for the remainder in integer division, however not the power operator. As in Python, the result of a mixed expression consisting of int and float is of type float. Unlike Python version 3, the '/' operation applied to arguments of type int denotes *integer division*.

Output

The output command printf can only contain string arguments. Numbers must be fitted to strings for output, similar to what we did with the format method in Python:

```
printf("%2d times %f is %.2f\n", 3, 4.2, 3*4.2);
```

In C the percent sign is used to identify placeholders. Note that the order of place-holders and values (and their type) must match.

Variables and Assignments

In Python we could assign any value of arbitrary type to a variable. In C that is not al-lowed. A variable must be *declared* to be of a particular type before use, for instance by statements of the form 'int a', or 'float b'. These declarations provide mem-ory space that can hold a value of the corresponding type. This static one-time, no longer changeable specification of memory requirement *before* program execution ultimately leads to more efficient programs.

However, the declaration does not yet give the variable any value. It must be *initialized* by a value assignment before first use. Assignments are identified by the operator '=' as in Python.

Declaration and initialization can also be combined into one statement, such as 'int a = 42'.

In general, assignments of new values can then be made as in Python, as long as type compatibility is respected.

A special feature of C is the notation a++ and ++a for a variable a of type int. By themselves, both abbreviations stand for the assignment 'a = a + 1' or equivalently 'a += 1'. The difference becomes apparent in assignments of the form 'b = a++' and 'b = ++a'. In the first case, the *current* value of a is assigned to the variable b, and then the value of a is incremented by 1. In the second case it is the other way round. First a is incremented, and then the *new value* is assigned to b. The same applies to a-- and --a.

6.2 Control Structures: Branches, Loops

As in every modern programming language, program execution can be controlled with conditional and repetition commands.

Conditional Execution

The basic form of a conditional statement in C is

```
if (<condition>) {<statements>} else {<statements>}
```

The identifiers in angle brackets are to be understood as place holders for actual conditions and statement blocks. A condition is an expression that evaluates to one of the Boolean values *true* or *false*. The curly braces enclose the statements to be executed when the condition is true or false. If a statement block consists of a single instruction, the braces are not required. The part 'else {}' can be omitted.

In C, any expression that evaluates to a number can be used as a condition. The value 0 means 'false', all others 'true'.

The comparison operators <, <=, ==, != return the value 1, if the relation is true, 0 otherwise. Conditions can again be composed with Boolean operators. In C, logical *and* is denoted by &&, logical *or* by ||, and the negation by a leading exclamation mark, as in the characterization of *not equal* by !=.

Loops

Like Python, C offers several types of iteration loops. The while loop is the easiest one. It is the same as in Python, only in C syntax:

```
while (<condition>) {<statements>}
```

The condition is again enclosed in parentheses, the statement block in curly braces. If a block consists only of a single statement, the braces are not required.

Example 6.2 We formulate the Collatz problem in Example 3.3 in the Python chapter as a C program. For the sake of brevity we only show the instructions *within* the main function. Of course, in an actual execution, the entire program must be present.

```
int n = 100;  // input
while (n > 1) {
    if (n % 2 == 0) n /= 2;  // integer division
    else n = 3*n + 1;
    printf ("%d\n", n); }
printf("arrived at 1\n");
```

There is also a for loop in C, but with a somewhat idiosyncratic form:

Example 6.3 The following program computes the reward for the chess inventor, explained in the introductory chapter. At first reading, it can be helpful to read 'int' for 'unsigned long'.

We declare and initialize the variables for field 1 by

```
1 int fieldno = 1;
2 unsigned long fieldval = 1;
3 unsigned long sum = 1;
```

Lines 1 through 3 set the number of grains for field 1 (i.e. 1 grain) and store it as the first contribution to the total.

The remaining computation is then performed in the loop:

```
4 for (fieldno = 2; fieldno <= 64; fieldno++) {
5     fieldval *= 2;
6     sum += fieldval; }
```

The entries in the loop header in line 4 are to be understood as follows:

The statement in the first entry is executed before entering the loop for the first time. Here the value 2 is assigned to the loop counter fieldno.

The second entry specifies that the loop is executed as long as the condition holds, here as long as fieldno has a value less than or equal to 64.

The third entry is executed at the end of each loop run, here denoting that the index of the field under consideration is incremented by 1.

The effect is that the loop is passed through once for every field ≥ 2 in the chessboard. The data type int generally has a word length of 4 bytes so that the large numbers required here cannot be stored. In many implementations of C, however, the type 'unsigned long' exists, which provides 8 bytes for unsigned numbers $n \geq 0$ up to $2^{64} - 1$. And that's exactly the maximum number we need.

We can then print the result 18446744073709551615 with

```
7 printf("%lu\n", sum);
```

Note the placeholder-identifier '%lu' for values of type 'unsigned long'.

In general, a for loop has the form

```
for (<initialization>; <condition>; <increment>) {<statements>}
```

The meaning of the components were explained in the example. Perhaps the general principle becomes even clearer, when we describe the effect with an equivalent construction based on a while loop:

```
<initialization>
while (<condition>) {
    <statements>
    <increment>}
```

Conversely, this construction gives rise to the question: what happens if the initialization and increment instructions in a for loop are both empty? Then, in fact, we get back the effect of a while loop. This is illustrated by an example.

Example 6.4 The sequence

$$a_0 = 1, \quad a_{n+1} = \frac{1}{2}(a_n + 2/a_n)$$

converges to $\sqrt{2}$. This allows us to approximate the root by

```
1 float r = 1;
2 float eps = 1.e-6;
3 for (; (r*r - 2 )*(r*r - 2) > eps*eps; )
4         r = (r + 2/r) / 2;
5 printf("%f\n", r);
```

The execution is stopped when $|w^2 - 2| \leq \epsilon$. This latter formulation is chosen so that we can work without the absolute value and the square root, neither of which are part of the basic C language.

The number 1.e-6 in line 2 is simply 0.000001 in exponential notation.

Note that we have declared the variables in lines 1 and 2 as float. We will generally continue to follow this standard convention. In practical applications it is usually advisable to use the more precise variant double instead.

Remark For specially interested readers: This hybrid form of the for loop as both counting loop, and general while loop, is not entirely satisfactory from the standpoint of computability theory. There a clear distinction is made between general loops and those, where the number of passes is known before the loop begins.

The class of so-called *primitive recursive* functions consists exactly of those functions, that can be computed using only *counting* loops. The larger class of μ-recursive functions also includes computable functions that require unlimited while loops. The prime example is the Ackermann function in Example 3.11 in the Python chapter.

6.3 Functions

We first consider functions over the basic types for numbers. A C function can take
a *tuple* as argument, but can return only a single value as output, all of which must
be labeled with their type. Otherwise the definition is similar to the one in Python.

Example 6.5 Here is the factorial function:

```
int factorial(int n) {  // declaration outside of main
    int res = 1;
    for (int i = 1; i <= n; i++) res *= i;
    return res; }
```

Note that the counter variable i is declared in the loop header. This is possible only
in modern C versions beginning with C99.

The function definition must be given *outside of the main function*. It is however
called *within* main, like so:

```
int a = factorial(5);   // use in main
printf("%d\n", a);
```

Remark When working with functions, it is important to understand how variables
and values are handled in the internal instruction block. The parameters of the function (in the example 'n') are pure placeholders, they assume the argument *value*. A
sequence of instructions 'int n = 3; int a = factorial(1);' returns the function
result 1, without effect on the outer variable n. The value of the argument is passed
to the function, not the variable itself.

Conversely, the variable res is introduced *within* the function definition and can
therefore only be accessed within that code block.

In fact, the scope of variables is a large topic, but we won't go into it here.

It should now be easy to reformulate all functions from the examples 3.8, 3.9, 3.10
and 3.11 in the basic Python chapter, including the recursive ones. For recursive functions, the above-mentioned fact is crucial that only values are passed to the functions,
not the variables themselves. What happens is that new variables are set up at each
recursion level to copy the values from the calling level.

Functions as Arguments

Each C function f has a specific *type*

```
<type> f (<type>, ..., <type>)
```

where the expression to the left of the function name denotes the return type and the
ones in parentheses to the right denote the types of the arguments.

This can be used to specify the type of a *function argument* in a higher-level function.

Example 6.6 We define a difference-quotient function diff_quot to approximate
the derivative f' of a given function $f: \mathbb{R} \to \mathbb{R}$ at a point x:

```
1 float diff_quot(float f (float), float x) {   // outside of main
2     float h = 1.e-6;
3     return (f(x+h) - f(x)) / h; }
```

The argument tuple in line 1 consists of a variable f of type 'float f (float)' and a 'float x'. In line 2 the value *h* needed for the approximate calculation of the derivative at the point x in line 3 is set, again to the value 1.e-6 = 0.000001 in decimal exponent notation.

To test it we apply diff_quot to the function

```
4 float f_ex(float x) { return 4*x*(1 - x); }   // outside of main
```

and the point $x = 0.5$:

```
5 printf("%f\n", diff_quot(f_ex, 0.5));        // in main
```

and get the result -0.000004.

Remark In the example we return the derivative function value at a specific point. Unlike Python, in C it is not directly possible to return the derivative *function* itself.

6.4 Arrays

C arrays are rather spartan (some even say primitive). An array always consists of a sequence of a fixed length of numbers of the same type.

For example, an array arr of length 3 with elements of type int is declared as 'int arr[3]'. The elements can be read and written as in Python lists. Note again that indices start at 0, hence the last element in the array of length 3 has index 2.

As with simple variables, the elements of an array must be initialized before use, such as:

```
int arr[3];                               // declaration
for (int i = 0; i < 3; i++) arr[i] = i;   // initialization
```

Note again that we have declared the variable i within the loop header. Since the array index runs from 0 to 2, it can only accept values less than 3, as specified by the condition in the second argument.

Another way to initialize is to assign the sequence of values as in line 1 below:

```
1 int arr[3] = {0, 1, 2};
2 int arr2[] = {0, 1, 2};
```

Line 2 shows an alternative method, in which declaration and initialization are combined. Here the compiler by itself recognizes that the array has a length of 3.

Because an array is of constant length and each entry is of the same known type, the compiler can allocate the required memory before the program run (in jargon: at compile time). If int has the word length 4 bytes, 12 bytes must be reserved for an int array of length 3. The byte length of a data type or a variable can be queried with the built-in function sizeof.

Remark Strictly speaking, `sizeof` returns the value not in bytes, but as a multiple of the shortest C data type `char`, which is usually however precisely 1 byte.

With `sizeof` you can conversely also regain the number of elements of an `int` array `arr` by dividing the memory requirement `sizeof(arr)` by that of its component type `sizeof(int)`. C itself does not provide a function like `len` in Python.

Remark The memory for an array of length n is thus determined before the program is started, but it is usually not checked during execution that the index limits are complied with. For example, if we make the assignment `arr[100] = 1` for our array `arr` above, a critical value may be overwritten, with serious consequences. Not for nothing, in the past you could sometimes hear: "C is a language for adults, who know what they are doing."

Arrays as Function Arguments

Arrays can be used as *function arguments*. However, you should also include the length as an additional argument when passing it, since calculations with `sizeof` usually do not work within a function. We will come back to this later.

Example 6.7 The following function calculates the Euclidean norm $||v|| = \sqrt{\sum_{i=1}^{n} x_i^2}$ of a vector $v = (x_1, \ldots, x_n) \in \mathbb{R}^n$. Note that the root function `sqrt` does not belong to the basic C language, but to the library `math`, which therefore has to be imported with `#include<math.h>` in the preamble. The functions in `math` are listed in the Wikipedia article `C_mathematical_functions`.

```
1 float norm(int n, float v[]) {
2     float s = 0;
3     for (int i = 0; i < n; i++) s += v[i]*v[i];
4     return sqrt(s); }
```

The function is called in the form `norm(n,v)` without square brackets:

```
5 float v[] = {0, 3, 4};
6 printf("%f\n", norm(3,v));
```

Note that in Linux the mathematical functions are bundled separately. To use them we have to pass the '-1m' option to the compiler. If the program is stored in a file, say, `norm.c`, it can be compiled and run as follows:

```
$ gcc norm.c -o norm -lm
$ ./norm
5.000000
```

Matrices

Matrices can be represented by 2-dimensional arrays, for example a 2×3 matrix A as follows:

```
int A[2][3];                      // declaration
for (int i = 0; i < 2; i++)   // initialization
    for (int j = 0; j < 3; j++) A[i][j] = 3*i + j;
```

Like linear arrays, matrices can also be initialized together with the declaration:

```
int B[2][3] = { {0, 1, 2}, {3, 4, 5} };
int C[][3]  = { {0, 1, 2}, {3, 4, 5} };
int D[][3]  = {0, 1, 2, 3, 4, 5};
```

In B and C the components are assigned row by row. In C the compiler itself determines the number 2 of rows from the number 6 of components and the number 3 of columns. The number of columns cannot be omitted because the compiler does not care about the inner curly braces, but treats the input as a linear array as in D. In D it is obvious that the number of columns is actually needed to split the array properly.

If you use using matrices as function arguments, you *must enter* the number of columns.

Example 6.8 We define a function to compute the determinant $\det(A)$ of an $n \times n$ matrix $A = (a_{ij})$ by expansion according to Laplace's formula, here along the first row:

$$(*) \qquad\qquad \det(A) = \sum_{j=1}^{n}(-1)^{1+j}a_{1,j} \cdot \det(A_{1,j}),$$

where each $A_{1,j}$ is the *minor* of A, obtained by erasing the first row and the jth column.

From $(*)$ we see that a *recursive* approach is appropriate:

```
 1  float det(int n, float A[][n]) {
 2      if (n==1) return A[0][0];              // recursion base case
 3      float sum = 0;
 4      for (int col = 0; col < n ; col++) {   // for each column:
 5          float A_sub[n-1][n-1];             // declaration of minor
 6          for (int i = 0; i < n-1; i++)      // def of entries
 7              for (int j = 0; j < n-1; j++)
 8                  A_sub[i][j] = j < col ? A[i+1][j] : A[i+1][j+1];
 9          float s = A[0][col] * det(n-1, A_sub);  // recursive Laplace-
10          sum += col % 2 == 0 ? s : -s; }         // expansion
11      return sum; }
```

In line 1 the number n of columns is passed to the function in the first argument, so that it is already available in the second argument, thus allowing a correct input of the matrix A. However, note that this evaluation order cannot generally be guaranteed. It works for the compiler used here, but it is not accepted by the C++ compiler in the next chapter.

In lines 5–8 the minor A_sub is constructed by erasing row 1 and column col in A.

Note that the expression on the right-hand side in line 8 is an abbreviated form of an if-else statement. The condition j < col between '=' and '?' is evaluated. If true, the value A[i+1][j] will be assigned to A_sub[i][j], otherwise the value A[i+1][j+1] following the colon.

In line 9 the determinant of the minor A_sub is computed by recursion, and then multiplied with the corresponding entry from the first row of the input matrix.

In line 10 the result s is added to – or subtracted from – the sum, as required by the expansion formula. Also note the use of the abbreviated if-else statement. The condition between '+=' and '?' is evaluated, and if true, the value of s is added to sum, otherwise that of -s.

The procedure in lines 5–10 is embedded in the outer for loop beginning in line 4, and hence applied to all columns.

We test the function det for the matrix A, given by:

```
12  float A[3][3] = { {1, 2, 3}, {1, 1, 1}, {3, 3, 1} };  // in main
13  printf("%f\n", det(3, A));
```

and obtain the correct result 2.000000. Note that C automatically converts the integer entries to floats.

6.5 Pointers

In the square-root computation with Python in Example 4.1 in the SciPy chapter, the solution *pair* was returned as a function value. How can we do that in C? As a matter of fact, we cannot do it directly. But we can specify a replacement construction, based on the most powerful (but also most dangerous) instrument of C, the *pointer*.

Let's start with the basics. When we declare a variable like 'int a', a suitable memory space is reserved. This area also has a storage location *address* which can be accessed directly with the *address operator* '&':
The instructions

```
int a;
printf("%p\n", &a);  // %p is the place holder symbol for pointer
```

return an address such as 0x7fff5fbff86c, which is simply a *hexadecimal* representation of the decimal number 140734799804524. The details should not bother us here.

Conversely, you can also introduce a variable that is used specifically to store the memory location of a number of type int.
This is declared as follows:

```
int* p;  // p can hold the address of an int number
```

The more abstract term *pointer* is usually used instead of 'address'. The pointer is then interpreted as a *reference* to the value. In truth, however, an address of the type mentioned above is always meant.

The actual value stored at the address can then be checked with the *asterisk-* or *dereference-operator* *p.

Thus the asterisk has a double meaning. In *declaration* it says that p is a variable of type int*, which means a pointer to int. In *use*, *p refers to the number value stored at that location.

Example 6.9 We want to write a function '`... val_pair(..., int x)`', so that after execution of

```
1 int a, b;  // possible: two declarations in same line
2 val_pair(..., 1);
```

the variable a contains the value 1, b the value -1.

The function has no return value, but performs the work behind the scenes, so to speak. In C this is indicated by the *return type* void.

As stated before, the function cannot assign a value to both a and b. But it *can* access the addresses, and change the *contents* there. The function call in line 2 will therefore have the form `val_pair(&a, &b, 1)`, where the first two arguments must be of the pointer type int*.

Here is the complete definition:

```
1 void val_pair(int* p, int* q, int x) { *p = x; *q = -x; }
```

The function `val_pair` can then be tested by

```
2 int a, b;  // in main
3 val_pair(&a, &b, 1);
4 printf("a = %d, b = %d\n", a, b);
```

Exercise Corresponding to the Python program in Example 4.1 in the SciPy chapter, write a C program to compute and output both square roots.

Arrays and Pointers

It is often said that C arrays are but slightly disguised pointers. There is certainly some truth to this. Internally, an array 'int a[n]' is managed like a pointer a of type int*. This can be verified as follows:

```
1 int a[] = {1, 2, 42};  // in main
2 int* p = a + 2;         // address arithmetic
3 printf("%d\n", *p);     // outputs 42
```

In line 1 a is *declared* to be of type array, and initialized with {1,2,42}.

In line 2 a is *interpreted* as an int pointer with the start address of the array as value, i.e. the address of the first element a[0]. The addition '+2' means: "add 2*sizeof(int)" to this address value in order to get a new address, namely that of a[2].

This address is then assigned to p and finally dereferenced by *p in order to get the int value 42 stored at the address.

The print command in line 3 returns that number.

An integer pointer only contains a specific address and only knows that 4 bytes are needed for an int, but not how many storage locations are required for the entire array. This is the reason why an array limit check cannot be easily performed.

This internal representation of arrays by pointers is also the reason why the query sizeof(a) within a function does not compute the size of the array, but the size of a pointer, which is always 8 bytes in 64-bit implementations.

Pointers as Arrays

The identification of arrays and pointers can also be reversed, and arrays declared as pointers. This is especially useful for large data sets, that are only needed in memory temporarily. Arrays are statically declared with a fixed size before program execution, while pointers can be dynamically requested ("allocated") while the program is running, and memory then released again, when it is no longer needed.

We won't go into the details here, but just indicate how arrays and matrices can be treated in this way.

Remark However, it should be emphasized, that the use of pointers, addresses and dynamic memory management are not C specific concepts, but ultimately form the basis in all modern programming languages. The only question is, how far the user is forced to deal with these technical mechanisms. This cannot be avoided in system programming, but in the application of computers to solve mathematical problems, it is certainly desirable, not to be confronted with such details.

Example 6.10 We want to declare an int array of length n as a pointer and allocate a corresponding memory area. For this, we need the malloc function (for memory allocation) from the library stdlib, which returns a memory space of a desired size. This is how it works:

```
#include <stdlib.h>              // in preamble
...
int* arr = malloc(sizeof(int) * n);  // in main
```

The result is an int pointer arr, which points to the beginning of a newly provided address block that can hold n int numbers. We can now use arr as a normal array, and access the elements with arr[i].

Example 6.11 For the same reason, matrices can also be created in this way. To illustrate, we consider a matrix of type 'int A[n][m]'. According to explanations above, each row can be viewed as an int pointer, the matrix therefore as an *array* of int pointers. With the same argument, the matrix can be identified again with a pointer to the beginning of this array. A could thus also be declared as being of type (int*)*, in C written as int**.

So, after a short breath, on to the actual implementation:

```
1 int** A = malloc(sizeof(int*) * n);  // in main
2 for (int i = 0; i < n; i++)  A[i] = malloc(sizeof(int) * m);
```

In line 1 the variable A is declared as a pointer to int pointers. A memory block to store n such int pointers is allocated, one for each row. Then A is set to point to the beginning of that block.

In line 2 a memory block is allocated to store m int values for each index i. Finally, the int pointer A[i] is set to point to the beginning of that block.

The matrix constructed in this way can now be used as usual, for example filled with values like this:

```
3 for (int i = 0; i < n; i++)
4     for (int j = 0; j < m; j++) A[i][j] = i*m + j;
```

When the matrix is no longer needed, its memory space can be released (or "deallo-cated") using the stdlib function free:

```
5 for (int i = 0; i < n; i++) free(A[i]);
6 free(A);
```

Note that memory must be released in reverse order of allocation.

6.6 Structures

Let's return once more to the square-root computation in Example 4.1 in the SciPy chapter, where the *pair* of *both* solutions could be returned as a function value.

In C we can use the *structure* concept to implement a similar behavior. We define an appropriate data type, which can contain *both* solution values, and can also be returned as result of a function.

A corresponding structure can, for instance, be introduced as follows:

```
1 struct xypair { int x; int y; };   // type definition
2 struct xypair p;                    // variable declaration
3 p.x = 1; p.y = 2;                   // initialization
```

In line 1 a data type 'struct xypair' is defined, which has two components, called *members*, x and y of type int. In line 2 a variable p of this type is declared. As seen in line 3, the members can be accessed by p.x and p.y.

It should be emphasized that the type is 'struct xypair' and not 'xypair'.

Instead of the full type name 'struct xypair' you *can* however introduce an *abbreviation* with the keyword 'typedef' as follows:

```
typedef struct xypair abr_xypair;
```

Now abr_xypair can be used as a synonym for 'struct xypair'.

It is even possible (and very common) to define a synonym using the original name, for example:

```
typedef struct xypair xypair;
```

Declaration and initialization can be combined also for structs. Using the abbreviation xypair for struct xypair we get at the following concise form, which corresponds to the original definition above:

```
xypair p = { .x = 1, .y = 2 };
```

Example 6.12 Based on the 'struct xypair' type, here denoted by the abbreviation xypair, we can then actually define a function xypair_fct that outputs a *value pair* of type xypair:

```
xypair xypair_fct(int x) {  // xypair is return type
    xypair p = { .x = x, .y = -x };
    return p; }
```

It should be noted that the type definition of 'struct xypair' and the typedef abbreviation must be made before the function declaration, i.e. in this case outside of the main function.

The function can then be used like this:

```
xypair p = xypair_fct(1);  // assignment to variable
int x = xypair_fct(1).x;  // access to component
```

Sparse Matrices

In Sect. 4.4 in the SciPy chapter we discussed sparse matrices and vectors.

Example 6.13 Here we show an example of how *sparse vectors* can be represented by linked lists consisting of structures and pointers.

A *linked list* consists of individual nodes, each containing data and a link to the next node.

We begin with a data type for the representation of nodes:

```
1 typedef struct node {
2     int index;
3     int value;
4     struct node* next;
5 } node;
```

This defines the data type 'struct node'. The keyword typedef declares that we want to call this type with the abbreviation written after the type definition, i.e. again with the designator 'node'.

The type node contains three entries: index will be used to store vector positions with nonzero values, value these values themselves. The pointer next serves as a *link* to the next index-value pair with a value $\neq 0$.

In order to use a node, we need to allocate memory for it. The following function create_node provides a pointer to a node to store index, value, and a pointer to the successor node. Note that here for the first time we encounter a function that returns a pointer.

```
6 node* create_node(int idx, int val) {
7     node* npt = malloc(sizeof(node));
8     (*npt).index = idx;
9     (*npt).value = val;
10     (*npt).next = NULL;
11     return npt; }
```

In line 7 memory is requested for a node structure, and npt declared as a pointer to it.

In lines 8–10 the external arguments idx and val are stored in the corresponding members of the node *npt. The 'next' member is as yet not linked to any successor. It receives the special value NULL. Formally, NULL is a pointer with a value 0. This can be verified with the instruction printf("%p\n", NULL).

Remark Notions like(*npt).index above are often written in a more concise form as npt->index. We follow that convention below.

As mentioned, we intend to represent vectors as linked lists of nodes. We now develop a suitable data structure.

The following data type 'struct vec', again abbreviated to vec, provides pointers to the first and last nodes.

```
12  typedef struct vec {
13      node* first;
14      node* last;
15  } vec;
```

The next function create_vector allocates memory for a vector of type vec and returns a pointer to it:

```
16  vec* create_vector() {
17      vec* v = malloc(sizeof(vec));
18      v->first = NULL;
19      v->last = NULL;
20      return v; }
```

Initially, the vector is empty. Lines 18 and 19 state that no nodes have yet been included in the vector representation.

The following function 'append' is used to actually create the linked list. Intuitively, it appends a node *npt to a vector *v. Technically it is again formulated in pointer terms:

```
21  void append(vec* v, node* npt) {
22      if (v->first == NULL) {
23          v->first = npt;
24          v->last = npt;}
25      else (v->last)->next = npt;
26      v->last = npt; }
```

The condition in the if-clause in line 22 is fulfilled if v has just been newly created by the function create_vector, hence as yet no node has been appended. In that case, the pointer to the node to be appended, i.e. npt, is assigned to *both* pointers first and last of *v.

Otherwise, *v has a last node, say tmp_last, to which *v.last points. In line 25 the new node *npt is linked to tmp_last as its successor.

In line 26 the node *npt is marked as the new last node in *v.

The following main program shows how a vector, given as a C array, is converted to a sparse vector.

Note that the program code considered so far belongs to the "preamble" between the #include statements

```
27 #include <stdio.h>
28 #include <stdlib.h>  // contains the malloc function
```

and the main function.

The following code must however be written *within* the main function:

```
29 int arr[] = {0, 2, 0, 0, 8, 10, 0, 14, 16, 18};
30 vec* v = create_vector();
31 for (int i = 0; i < 10; i++) {
32     if (arr[i] != 0) {
33         node* npt = create_node(i, arr[i]);
34         append(v, npt); } }
```

In line 29 we fix a test array arr with 10 elements.

In line 30 the (initially empty) vector vec is created, and then populated in the for loop in lines 31–34. The condition in line 32 checks whether the arr entry has a value $\neq 0$, hence should be included in the linked-list representation. If so, a new node containing its index and value is created in line 33, and appended to vec in line 34.

The construction can be verified as follows:

```
35 for (node* npt = v->first; npt != NULL; npt = npt->next)
36         printf("%d %d\n", npt->index, npt->value);
```

In the for loop in lines 35 and 36, all nodes of the sparse vector are run through and their index-value pairs displayed. It is worthwhile to follow how the process is controlled by the entries in the loop header in line 35.

6.7 Files, Input and Output

As a simple example we show how a matrix can be written to a file and read out again.

Example 6.14 Consider the matrix

```
1 float A[2][3] = { {1., 2., 3.}, {4., 5, 6.} };
```

For writing we use the fprintf statement:

```
2 FILE* fp = fopen("test.txt", "w");
3 int n = 2, m = 3;
4 fprintf(fp, "%d %d ", n, m);
5 for (int i = 0; i < n; i++)
6     for (int j = 0; j < m; j++)
7             fprintf(fp, "%f ", A[i][j]);
8 fclose(fp);
```

Line 2 declares a pointer fp to an object of the C data type FILE, and links it to the file test.txt in write mode with the command fopen.

In line 3 the row and column numbers of the matrix A are parametrized, so that the program can also be used for other values. In line 4 these values are written to the beginning of the file.

The instruction fprintf works like printf, with the exception of the additional *pointer* that directs the output to the file.

In line 7 fprintf writes the matrix elements to the file.

Line 8 terminates the connection to the file.

In the following program the file content is read back into a matrix B. For this we use the counterpart to fprintf, namely fscanf:

```
1 FILE* fp = fopen("test.txt", "r");
2 int n, m;
3 fscanf(fp, "%d %d ", &n, &m);
4 float B[n][m];
5 for (int i = 0; i < n; i++)
6     for (int j = 0; j < m; j++)
7         fscanf(fp, "%f ", &B[i][j]);
8 fclose(fp);
```

Line 1 opens the file in read mode. In line 3 the row and column numbers of the matrix are loaded into the variables n and m. In line 4 a matrix B is declared, and in line 7 populated with the values from the file.

The following for loop shows that the matrix has been completely reconstructed:

```
9  for (int i = 0; i < n; i++) {
10     for (int j = 0; j < m; j++)
11         printf("%f ", B[i][j]);
12     printf("\n"); }
```

Remark In line 4 a new $n \times m$ matrix is declared, for which the values of n and m are not yet known at program-compile time. This is actually allowed in C version C99. In previous versions you would yourself have to take care of allocating the required memory at runtime.

6.8 Conclusion

C is an extremely powerful language when you are prepared to dive into machine-specific internals such as memory management and addressing.

What is missing, however, are higher-level concepts such as dynamically change-able lists, or functions, whose return value is not restricted to basic data types.

Many of these concepts, which we saw earlier in Python, are provided in the C++ language to expand C's inventory. This is explained in the next chapter.

Chapter 7
The C++ Language

As the name suggests, C ++ was designed as an extension of C. The language was developed in 1979 by the Danish mathematician and computer scientist Bjarne Stroustrup as "C with classes". The model for the central concept of class construction was the Simula language, which was developed in Norway in the 1960s. The tutorial at cplusplus.com is recommended for an introduction.

The development environments already mentioned for C can be used also for C++. All examples in this chapter were tested in Xcode under macOS for the C++ variant C++14.

In Ubuntu Linux you can store programs in files with the extension '.cpp', and then compile them with 'g++'.

7.1 Transfer from C

Essentially, C++ comprises the whole language scope of C, so that all basic concepts considered in the last chapter can also be used in C++. In particular, this applies to the *number types*, *variables*, *control structures*, and *functions*. Also the *pointer constructions* and the C *structures* are available without limitation.

The development of the two languages is not always completely synchronous, so that occasionally some deviations occur. This concerns for instance the use of arrays as function arguments. The recursive determinant function defined in Example 6.8 is not accepted by the C++ compiler used here.

Be that as it may, C++ provides much more user friendly alternatives, so one should rather refrain from using C arrays.

Let's get straight on to the welcome example.

Example 7.1 The following is a perfectly correct C++ program:

```
#include <iostream>
int main() { printf("Hello World\n"); }
```

© Springer Nature Switzerland AG 2020
E. Smith, *Introduction to the Tools of Scientific Computing*, Texts in Computational
Science and Engineering 25, https://doi.org/10.1007/978-3-030-60808-8_7

The only thing that stands out is that the printf command has been moved from the library stdio to another one iostream (and that the extension '.h' is missing).

However, using printf is not considered good C++ style. In the sense of "C with classes", C++ prefers the "object-oriented" approach also for input/output.

7.2 Basics

We turn to some C++ specific basics.

The Output Operator <<, Input/Output to Files

Example 7.2 In proper C++ style the welcome example looks like this:

```
#include <iostream>
int main() { std::cout << "Hello World\n"; }
```

The object cout controls the output to the screen. (See below for the meaning of the additional prefix 'std::'.) It takes the expression to the right of the *output operator* '<<', processes it, and passes the result to the output device. In our example, cout recognizes that a string of characters should be written, followed by a line break.

It is also possible to pass several expressions in succession, even consisting of different data types. For example, the input

```
std::cout << 1./3 << " times " << 3.14
          << " is " << 1./3*3.14 << std::endl;
```

gives the result '0.333333 times 3.14 is 1.04667'. Even the data types are automatically recognized by cout, without need for type placeholders such as '%f' in printf.

The use of std::endl (for end line) is essentially identical to that of '\n'.

Example 7.3 To format the output, appropriate parameters can be set in cout. For example, consider the input

```
1 float p = 3.14159;
2 std::cout.precision(4);            // 4 digit precision ...
3 std::cout << p << "\n";            // Out: 3.142
4 std::cout.setf(std::cout.fixed);   // ... here for the decimals
5 std::cout << p << "\n";            // Out: 3.1416
6 std::cout.width(8);
7 std::cout << p << "\n";            // Out:    3.1416
```

In line 2 the output accuracy is set to 4 digits. In line 3 the number *p*, rounded to 4 digits, is printed.

The method setf (for set flag) in line 4 can control various yes/no-switches, here specifically, that the output accuracy 4 should refer to the decimal places.

Line 6 states that the output is to be printed right aligned within a space of width 8. Note that the `cout` settings from lines 2 and 4 are still valid. They remain valid until changed again.

Writing to Files

Writing to a file is quite similar, except that it is not controlled by `std::cout`, but by a file manager imported by

```
#include <fstream>
```

The following code is again part of the `main` function:

```
1 std::ofstream myfile;  // 'o f' for output file
2 myfile.open("example.txt");
3 myfile << "Text in file\n";
4 myfile.close();
```

In line 1 the type declaration 'ofstream myfile' indicates that the following output data *stream* should be routed to a file that is referenced by the `myfile` object.

Line 2 establishes the connection to the actual external file.

In line 3 text is written to the file. Note that we could also redirect all formatted output in Example 7.3 by simply replacing the target `std::cout` with `myfile`.

Line 6 closes the file connection. The instruction is not mandatory; at the latest, at the end of the program this happens automatically.

Reading from Files

There are many options for reading from files. We look at a simple variant that uses the function `getline` to read a text file, line by line, and output it to the screen:

```
1 std::ifstream myfile;  // 'i f' for input file
2 myfile.open ("example.txt");
3 std::string line;
4 while (!myfile.eof()) {
5     getline(myfile, line);
6     std::cout << line << '\n'; }
```

In line 1 a variable `myfile` of type `ifstream` is declared, and in line 2 linked to the file to be read.

Line 3 declares a variable 'line' of type `string`, which in line 5 receives the next line of the file `myfile` from the function `getline`, and outputs it to the screen in line 6.

The condition in the loop head in line 4 states that the loop is repeated as long as the end of the file has not yet been reached. The method `eof` (for: end of file) returns the Boolean value 'true' when the end is detected, the exclamation mark negates this condition, so that the loop is executed as long as the end is *not* reached.

Remark The data type string does not belong to the C++ core language, and therefore should be imported through '#include <string>'. But at least for C++14 under Xcode on the Mac, this is not necessary. There, string is also available without explicit import.

Namespaces

The prefix 'std::' before identifiers like cout serves to avoid name collisions. This is quite analogous to the convention that we already met in Python. For example, the function sin appears in both packages math and numpy. If both are imported, we can only distinguish which one is meant by a prefix like math.sin or numpy.sin. Correspondingly, std contains a list of names in which all identifiers from the standard libraries are collected, so that no ambiguity will arise.

If you are confident that the identifiers do not lead to conflicts, you can issue the statement 'using namespace std'. You can now write cout instead of std::cout. In the following we will adhere to this convention, and use identifiers without "origin marking".

7.3 Lambda Expressions

Unlike C, in Python a function could return another function as output. Recent versions of C++ (from C++11 onwards) now provide *lambda expressions*, which at least in simple cases also allow such constructions. As an example, we show how a second level function ddx is defined, which returns the (approximate) derivative function ddx(f) of a function f.

But first to the lambda expressions themselves. A simple example:

```
[](int n) { return n*n; };
```

This defines the function $n \mapsto n^2$. Here the bracket pair [] is the identifier that tells the compiler that the definition of an anonymous function follows. The arguments are enclosed in parentheses. The code block defining the function is enclosed in braces. In simple expressions as the one considered here, there is no need to declare the type of the return value; the compiler takes care of that.

A lambda expression can then be applied to arguments directly, or be assigned to a variable. In our example, the assignment to a function variable fct looks like this:

```
function<int(int)> fct = [](int n) { return n*n; };
```

The term 'function<int(int)>' denotes the C++ type declaration for functions from int to int.

Note that in C++, functions can either be declared in the preamble or, in contrast to C, in main.

Now `fct` can be used as a normal function to compute, say, `fct(4)` with the output 16.

Actually, the lambda construction is even more powerful. It can reference variables from the environment, in which it is defined, in jargon: *capture variables*. This is best explained when we now turn to the actual implementation of our differential operator `ddx`.

In order not to be distracted by rounding errors, we use double precision floating point numbers.

```
1  function<double(double)> ddx(double f (double)) {   // in preamble
2      function<double(double)> f_prime;
3      f_prime = [f](double x){ double h = 1e-6;
4                              return (f(x+h) - f(x)) / h; };
5      return f_prime; };
```

In line 1 the *argument* of `ddx` is declared as a function variable `f` which takes a `double` as argument, and also returns a `double` as value. This is normal C notation, already encountered in Example 6.6 in the C chapter.

The *return value* of `ddx` is of essentially the same type, now however declared as a genuine C++ type 'function<double(double)>'.

In line 2 a variable `f_prime` of that type 'function<double(double)>' is declared and then initialized by the lambda expression in lines 3 and 4.

Note that line 4 refers to the function `f`. As stated above, this is possible because in line 3 the variable `f` is "captured" in the lambda expression. The bracket pair [ ] serves as capture-operator.

We test `ddx` for the function $g : x \mapsto x^2$:

```
double g(double x) { return x*x; };      // in preamble
...
function<double(double)> dgdx = ddx(g);  // in main
cout << dgdx(2) << endl;   // Out: 4
```

Note that the output is 3.8147 if we use `float` instead of `double` precision.

7.4 Data Type vector

With the data type `vector`, C++ offers a user-friendly alternative to C arrays. C++ vectors do not decay into pointers, the number of components can always be queried directly, they can be used as function arguments without restriction. In addition, memory management is organized by the system itself. The data type is provided by the statement '#include <vector>' in the preamble.

There are several options for declaration and initialization, such as:

```
1  vector<int> u(3);   // 0-initialized
2  vector<int> v(3,1);
3  vector<int> w = {1, 2, 3};
```

Here we again assume 'using namespace std' to be declared. Otherwise, we would have to use the longer form std::vector.

The instruction in line 1 defines a vector u, which consists of 3 elements of type int. All components are implicitly initialized with a 0 value. Line 2 is analogous, except that now all elements are explicitly initialized with the value 1. In line 3 the vector w is implicitly declared to be of length 3 and initialized with the values enclosed in braces.

As in C arrays, components are accessed in the form v[i] etc. The length of a vector v can be inquired with v.size(). However, note that as in C, range bounds are not controlled.

Very useful is the ability to dynamically adjust the length of a vector. For example, the statement 'vector<int> v' first sets up an "empty" vector of length 0, which is then converted to a vector of, say, length 3 by v.resize(3), or by v.resize(3,1), where then additionally the value 1 is assigned to all components.

Sparse Vectors

In Sect. 6.6 in the last chapter we saw how to represent sparse vectors by linked lists in C. Using C++ vectors, the construction can be significantly simplified.

We show how to convert a C array

```
1 int arr[] = {0, 2, 0, 0, 8, 10, 0, 14, 16, 18};
```

to a C++ representation for sparse vectors.

First we introduce a struct to represent index-value pairs:

```
2 typedef struct node {  // no pointer to successor node required
3       int index;
4       int value;
5 } node;
```

We then prepare an initially empty node vector

```
6 vector<node> node_vec;  // empty vector for components of type node
```

Finally, all index-value pairs of arr with value ≠ 0 are appended to vec one after the other:

```
7 for (int i = 0; i < 10; i++) {
8     if (arr[i] == 0) continue;              // proceed if != 0
9     node nd = { .index = i, .value = arr[i] };  // produce node
10    node_vec.push_back(nd); }               // append to vec
```

To verify our construction, we check whether all arr components have been transferred correctly. We take this as an opportunity to explain the C++ *iterator*:

```
11 vector<node>::iterator it;
12 for (it = node_vec.begin(); it != node_vec.end(); it++)
13     cout << ' ' << '(' << it->index << "," << it->value << ")";
14 cout << endl;
```

In line 11 the variable it is declared as iterator for the type vector<node>, comparable to a conventional number index in for loops. In line 12 the iterator it is set to the first element of the vector node_vec. Then the loop is executed repeatedly for each node of the vector, indicated by the third argument it++ in the loop head. This continues until the iterator has reached the last element indicated by the second argument in the loop head.

The expressions 'it->index' and 'it->value' in line 13 reveal that C++ iterators are based on the idea of pointers.

Remark In addition to the vector type, C++ also offers various other so-called sequential *container* types, i.e. data types that aggregate data. The type list, for example, in contrast to vector, enables elements to be inserted at any positions.

Matrices

A minor disadvantage is that matrices cannot be defined directly as two-dimensional vectors. Instead a matrix is constructed as a vector, where the entries are again vectors. A 3×2 matrix A of integers can for instance be defined like this:

```
vector<vector<int>> A(3, vector<int>(2));
```

Here A is declared as a vector of length 3, with each component being an int vector of length 2.

The matrix elements are again accessed by A[i][j]. As in vectors, all entries are by default initialized with the value 0.

It is also possible to explicitly pass the initial value 1 to all elements, simply by replacing '(2)' with '(2,1)' in the matrix definition above.

7.5 Reference Operator

One of the advantages of C++ over C is that C++ significantly reduces the need for low-level constructs such as pointers. However, sometimes it is useful to specify a variable b that can access the same memory space that belongs to another variable a. In C++ this can be achieved with the so-called *reference operator* '&'.

As a reminder, if we assign the address &a of a variable a to a pointer p, the contents of *p and a are linked together, so that a change in either will also automatically affect the other, precisely because both identifiers refer to the same memory space.

In contrast, when assigning b = a between variables of a value type, only the *current* value is *copied*, so that a subsequent change is not transferred from both to the other.

The *reference operator* '&' now allows the above tight coupling also directly between variables.

If, for instance, a denotes an int variable, then the statement

```
int& b = a;
```

declares b as a variable of type int, whose *address* &b *coincides* with the one &a of a.

The identifier b is now, so to speak, *synonymous* with a, or in C++ jargon, a *reference* to the variable a.

The reference operator will be of great use in the next section. It will permit us to assign a value to a component a[i] of an array, without explicitly referring to the index i. Here is the idea:

Example 7.4 The variable ref is to be used as a reference to the component a[0] of an array a:

```
int a[] = {1, 2};
int& ref = a[0];
```

If now the value of ref is incremented with ref++, then a is changed to {2,2}.

7.6 Classes

In the Python chapter, we explained the basic ideas in object-oriented programming, illustrated with a class Fraction. We show here, how the same example, a class for representing and managing fractions of integers, can be set up in C++.

Example 7.5 (A Class Fraction) The following specification can be written either in the preamble or in the function main.

```
1  class Fraction {    // in preamble or main
2      public:
3          int num, den;
4          Fraction(int numerator, int denominator)  {
5              num = numerator; den = denominator; }
6          Fraction add(Fraction b) {
7              return Fraction(num*b.den + b.num*den, den*b.den); }
8          bool operator==(Fraction b) {
9              return num*b.den == den*b.num; }};
```

First, we only consider the part that consists of lines 1–5. In line 1 the key word class declares that the definition of a class follows, here named Fraction. The keyword 'public' in line 2 states that all variables and functions defined in the following block can be accessed also from outside of the class instance. In line 3 two internal variables are declared, which take the numerator and denominator of the fraction to be represented.

Lines 4 and 5 specify a so-called *constructor*. It controls what happens, when a new class instance is set up. (This corresponds to the __init__ method in Python.) Note that the constructor is a distinguished function, recognizable by bearing the name of the class, and having no return value.

A new instance of the class Fraction is generated by a statement of the form

```
Fraction a(1, 2);   // class name, identifier a, initial values
```

The constructor reads the initial values 1 and 2 into its placeholders numerator and denominator, and in line 5 assigns them to the internal variables num and den.

Lines 6 and 7 define a *method* add to add fractions in Fraction representation. As explained in the corresponding Python example, it is called in the form a.add(b) for objects a, b of class Fraction. Similar to Python, we can make the sum notation available in infix form a + b. In C++ only the identifier add in line 6 has to be replaced by 'operator+'. Various operators can be overloaded in this way.

In lines 8 and 9 we find another example. Here the '==' operator is overloaded, so that fractions can be tested for equality in the form 'a==b':

```
Fraction b(2, 3);
cout << (a + b == b + a) << endl;
```

The value 1 is output, which is here to be read as the Boolean value true.

A Class Matrix

Let's look at a slightly larger example, a class for processing matrices.

The first question that arises is how matrices should be represented. Here we choose the approach outlined above, to consider matrices as row vectors of column vectors. We can then get into the details right away.

Here is a first draft:

```
1  class Matrix {
2          vector<vector<float>> mat;   // empty matrix
3      public:
4          int rows, cols;              // internal row/column numbers
5          Matrix(int r, int c) {       // constructor
6              rows = r; cols = c;
7              mat.resize(rows, vector<float>(cols, 0)); }
8  };   // end of class def
```

Line 2 creates an empty matrix mat as a vector of vectors. In line 6 the constructor assigns the number of rows r and columns c of mat to the internal variables rows and cols.

In line 7 the internal variable mat is converted to the correct shape, and initialized with 0-entries. (As indicated above, the explicit initialization with 0 is actually redundant, but it is always good practice not to depend on "hidden" features.)

The class definition is now already sufficient to declare Matrix objects. An $n \times m$ matrix A will be instantiated in the form

```
Matrix A(n,m);
```

The number of rows and columns can be queried by A.rows and A.cols.

However, we do not yet have access to the matrix elements. The variable mat is declared *outside* of the public area, and can therefore not be addressed directly in the form A.mat.

So we need a public access method. We define a variant that is not the technically most simple, but one that would suit a mathematical user.

Namely, we will be able to access the elements of the matrix A in the form

$$A(i, j) \quad \text{for} \quad 1 \le i \le n, \ 1 \le j \le m,$$

with indices starting at 1.

For this purpose we have to shift all indices by -1 in order to access the corresponding values in the internal variable mat.

This is achieved by including the following method in the public area of the class, for instance inserted after line 7:

```
float& operator() (int i, int j) { return mat[i-1][j-1]; }
```

In the return statement, the internal mat values are exported. To understand what happens, assume for a moment that the function declaration begins with, say, the declaration 'float get' instead of 'float& operator()'. Then the values would be accessible in the usual A.get(i,j) notation.

However, the function here actually has the unusual name 'operator()'. This means that the usage of the *pair of parentheses* '()' is *overloaded*, such that access to the return values becomes possible using the functional notation A(i,j). (Compare the corresponding construct 'p(x)' for the polynomial class in Sect. 3.9 in the Python chapter.)

But this only takes care of *reading* access. In comes the reference operator '&', appended to the float type of the return value. As indicated above, the declaration as 'float&' has the effect that A(i,j) has the *same address* as mat[i-1][j-1], hence we can access that internal storage space also in *writing* mode. (Recall that indices in the internal representation are always 1 less than in the external matrix object.)

The following method can be included to print a matrix in the usual row-by-row form:

```
void printm() {
    for (int i = 0; i < rows; i++) {
        for (int j = 0; j < cols;  j++) cout << mat[i][j] << "\t";
        cout << endl; }}
```

The control character '\t' causes a jump to the next tabulator position.

An addition method can be defined along the usual lines:

```
1 Matrix operator+(Matrix B) {
2     Matrix C(rows, cols);  // internal var for sum matrix
3     for (int i = 1; i <= rows; i++)
4         for (int j = 1; j <= cols; j++)
5             C(i,j) = (*this)(i,j) + B(i,j);
6     return C; }
```

In line 2 a new matrix C is instantiated. In line 5 it is populated with the correct sum. Note the expression '*this'. It stands as placeholder for the object that calls the method. (This corresponds to the use of the identifier self in Python, see Sect. 3.9 in the Python chapter.) It's sort of the C++ identifier for "me". Technically, after instantiation of the object, the *pointer* this contains its address.

The methods defined so far can be tested by

```
Matrix A(2, 3);
(A + A).printm();
```

We want to supplement our class skeleton with the *determinant* function det. To illustrate the possibilities, we choose to formulate it as a standalone function, not as a class method.

The program is based on the Laplace expansion in Example 6.8 in the C chapter.

```
1  float det(Matrix A) {
2      int n = A.rows;
3      if (n != A.cols) {          //  if not square: Abort
4          cerr << "Error\n";
5          exit(41); }
6      if (n == 1) return A(1,1);  // recursion's base case
7      Matrix B(n-1,n-1);          // declaration of minor submatrix
8      float d = 0;
9      for (int c = 1; c <= n; c++) {
10         for (int i = 1; i <= n-1; i++)
11             for (int j = 1; j <= n-1; j++)
12                 B(i,j) = j < c ? A(i+1,j) : A(i+1,j+1);
13         float s = A(1,c)*det(B);
14         d += c % 2 == 0 ? -s : s; }
15     return d; }
```

In lines 3–5 the computation is aborted with an error message, if the input matrix is not square. The word "Error" is written to the standard output for error messages cerr (err for error), which is often just the same screen used for cout. Then the program execution is aborted and the message: "Program ended with exit code: 41" is returned.

Declaration Order

The function det refers to the type Matrix, hence can only be defined *after* that type has been introduced. Since function definitions must be made outside of the main function, the only possible sequence here is: class Matrix - function det - function main.

However, it is clear that large programs will find it difficult or even impossible to meet such strict dependency requirements.

Example 7.6 For example, consider two functions f and g that recursively call each other, like:

```
2 int f(int n) { return n < 0 ? 22 : g(n); }
3 int g(int n) { return f(n-1); };
```

The standard way to deal with such dependencies is to follow the *forward declaration* concept. In the example it is sufficient to insert the new line 1 before line 2:

```
1 int g(int n);
```

In line 1 the function g is *declared* to be of correct type for use in line 2, even if the compiler at that point does not know what g does. However, the declaration is sufficient to interrupt the dependency cycle.

7.7 Header Files

In C++, the separation between declaration and implementation seen in Example 7.6 is omnipresent. Especially with large class definitions, it is common to collect declarations and implementations in separate but associated files.

As a typical example we consider our above class `Matrix`.

First we define the so-called *header* file `matrix.hpp`, which contains the class declaration, the variables used and the declarations for the methods and functions, but not their implementations:

```
1 #include <vector>
2 using namespace std;
3 class Matrix {
4     vector<vector<float>> mat;        // empty Matrix
5 public:
6     int rows, cols;                   // internal vars
7     Matrix(int r, int c);             // constructor
8     void printm();                    // remaining methods
9     float& operator() (int i, int j); //    dito
10    Matrix operator+(Matrix B); };    //    dito
11 float det(Matrix A);                 // def outside of class
```

For all components it is now clear, *how* they are applied, but not yet *what* they do. Note that the function `det` is again declared outside of the `Matrix` definition.

Also note also that the file has the extension `hpp`, which is common but not mandatory.

We come to the accompanying implementation file `matrix.cpp`. It collects the same definitions as before, so we ask the reader to copy and paste them from the pages above:

```
1 #include "matrix.hpp"  // imports all declarations
2 #include <iostream>    // needed for the code in printm below
3 Matrix::Matrix(int r, int c) { /* code as  before */ }
4 void Matrix::printm() { /* code as  before, needs iostream */ }
```

```
5 float& Matrix::operator() (int i, int j) { /* code as  before */ }
6 Matrix Matrix::operator+(Matrix B) { /* code as  before */ }
7 float det(Matrix A) { /* code as  before */ }
```

In line 1 the whole content of the header file is made available.

Note that all class methods must be prefixed with the class name 'Matrix::'. This does not apply to the det function in line 7, since det is not defined *within* the class.

All definitions can now be used within the function main in the file matrixmain.cpp:

```
#include <iostream>
#include "matrix.hpp"  // imports all definitions and implementations
int main() {
    Matrix A(2, 2);
    A(1, 1) = 1;
    A(2, 2) = 1;
    (A + A).printm();
    cout << det(A) << endl; }
```

Compiling and Executing the Program

If the program was created in a dedicated development environment like Xcode, the connection between the files is established automatically. The program can then be built and executed by clicking the Run button.

To compile and run the program under Ubuntu Linux, the easiest way is to store the files matrixmain.cpp, matrix.hpp, and matrix.cpp in the home directory and then compile both cpp files:

```
$ g++ matrixmain.cpp matrix.cpp -o matrixtest
```

to an executable program matrixtest, which can now be run as usual:

```
$ .\matrixtest
```

Remark As a final note on C++ classes for linear algebra: There is a vast number of ready-made classes available, both commercial, but also as open source. A popular choice is, for instance, the package *Eigen*, available on eigen.tuxfamily.org.

7.8 Summary and Outlook

In the last few chapters we discussed the basics of the Python, C and C++ programming languages, focussing on the concepts that are important for scientific computing. However, all three were developed as *universal* languages that are not specifically geared towards mathematical computation.

If you're mainly interested in fast number processing, then the Python approach of type-less variables, which comes with expensive internal memory management, may be a bit too liberal. Moreover, in Python the interpretative command execution tends to lead to longer run times. To some extent, these drawbacks can be overcome by using special types of data structures such as NumPy arrays. Also on the positive side, many specific mathematical libraries have been developed for the Python language, which turn it into a comprehensive user-friendly mathematical development environment.

C is the other extreme, so to speak. The tight machine oriented approach makes it possible to create highly efficient programs, efficient in terms of both runtime and storage utilization. However, this comes at the expense of less user-friendliness and less control of program security. In particular, the generous use of pointers is often seen as an uncertainty factor.

High-level programming languages such as C++ attempt to combine the efficiency of C with problem-tuned, more controllable data structures, for instance vectors and classes discussed in this chapter.

One approach to mathematical programming is to develop programs in a user-friendly language, such as Python, and then, if necessary, replace resource-intensive parts with modules formulated in efficiency-optimized machine-oriented languages. We will see examples later in the FEniCS chapter.

Another approach is to use languages that are inherently designed for mathematical programming and combine the ease of use of Python with the performance advantages of C. In the next chapter, we will discuss a particularly promising newcomer in this field, the Julia language.

7.9 Exercises

Numerical Integration, Simpson's Rule

Simpson's rule for the numerical approximation of definite integrals is based on

$$\int_a^b f(x)\,dx \approx \frac{b-a}{6} \cdot \left(f(a) + 4f\left(\frac{a+b}{2}\right) + f(b) \right).$$

Exercise 7.1

(1) Similar to Exercise 3.7 in the Python chapter, write a C/C++ function to determine the integral of a function f over an interval $[a,b]$, but this time using the *composite Simpson's rule* for n subintervals.

(2) Use it to compute the integral $\int_0^\pi \sin(x)\,dx$, such that the result is correct up to 6 digits.

Note In numerical computations with C/C++, it is advisable to use the type `double` instead of `float`. Most standard mathematical functions including sin are defined in `math.h` (in C++ use '#include <cmath>').

Polynomial Interpolation

Exercise 7.2 Write a program to compute a polynomial

$$p(x) = \sum_{k=0}^{n} a_k x^k$$

that interpolates the $(n+1)$ points $(x_0 . y_0), (x_1 . y_1), \dots, (x_n . y_n)$, based on the so-called *Vandermonde matrix* $V = (x_i^j)_{0 \le i, j \le n}$.

Test the program for $y_i = \tan(x_i)$ on the evaluation points $-1.5, -0.75, 0, 0.75, 1.5$.

Note This exercise is for "educational purposes only". In practice, the method is computationally too costly.

Ordinary Differential Equations

Exercise 7.3 Consider the boundary value problem

$$u'' - 4u = 0,$$
$$u(0) = 0, \; u(1) = e^2 - e^{-2}$$

in the interval $[0, 1]$. Solve this BVP numerically, using the method of discrete differences, by computing the corresponding difference equations on 5 equidistant points.

Exercise 7.4 Similarly, solve the BVP

$$u'' - 5u' + 4u = x^2,$$
$$u(0) = 0, \; u'(1) = 0$$

in the interval $[0, 1]$ by considering a corresponding difference equation on 5 interval points with distance $h = 0.25$, adding an additional point $1 + h$ to the right of the interval, in order to approximate the Neumann boundary condition. Compare your solution to the analytical solution

$$u(x) = -0.659305 e^x + 0.0030549 e^{4x} + \frac{1}{4} x^2 + \frac{5}{8} x + \frac{21}{32}.$$

Classes

The class definitions in the following exercises should follow the pattern above, namely to split declaration and implementation into separate files.

Polynomials

Exercise 7.5 Develop a class for the representation of polynomials according to the model provided in Example 3.19 in the Python chapter. Subclassing may be used, even if it was not covered in the text, but this is of course not mandatory.

Quaternions

Adding two more imaginary units j, k to the one i in the complex numbers, and defining multiplication between these units by a formula carved in stone (literally, cut on a stone of Broome Bridge in Dublin, Ireland):

$$i^2 = j^2 = k^2 = ijk = -1,$$

we arrive at the so-called *quaternions*.

The set of quaternions can be identified with the vector space $\mathbb{R}^4$ by assigning

$$i := (0, 1, 0, 0), \quad j := (0, 0, 1, 0), \quad k := (0, 0, 0, 1).$$

Multiplication of two quaternions $x = (x_1, x_2, x_3, x_4)$ and $y = (y_1, y_2, y_3, y_4)$ then looks like this:

$$x \cdot y = (x_1 y_1 - x_2 y_2 - x_3 y_3 - x_4 y_4,$$
$$x_1 y_2 + x_2 y_1 + x_3 y_4 - x_4 y_3,$$
$$x_1 y_3 - x_2 y_4 + x_3 y_1 + x_4 y_2,$$
$$x_1 y_4 + x_2 y_3 - x_3 y_2 + x_4 y_1).$$

With this multiplication, the quaternions form a *division ring* $\mathbb{H} = (\mathbb{R}^4, +, \cdot, 0, 1)$, also called a *skew field*, where every $x \neq 0$ has an inverse x^{-1} such that $x^{-1} \cdot x = x \cdot x^{-1} = 1$, but multiplication in general is not commutative.

Exercise 7.6 Develop a class for the representation of quaternions.

It should at least provide the additive and multiplicative constants 0 and 1, addition, subtraction, multiplication and division in infix notation, and the multiplicative inverse for $x \neq 0$.

A test for equality should also be possible using infix notation. Here two quaternions should be considered equal, if all components are indiscernible within `double` precision.

Illustrate with suitable examples.

Chapter 8
Julia

Julia is a newer programming language, specifically designed to address the needs of high-performance numerical analysis and computational science. The language intends to combine the efficiency of C/C++ with the convenience of Python. Work on Julia was started in 2009 by Jeff Bezanson, Stefan Karpinski, Viral B. Shah, and Alan Edelman at the Massachusetts Institute of Technology. The first final version 1.0 was released on 8 August 2018. There is apparently no official reason for the name "Julia".

The easiest way to learn and experiment with Julia is to download the installation from the official site `julialang.org`. It is available for free for Windows, Mac or Linux. That site also offers excellent comprehensive documentation on all aspects of the language. A good tutorial can be found on `en.wikibooks.org/wiki/Introduc-ing_Julia`.

The book [1], originally based on MATLAB, can also be used to learn Julia. Julia versions of the source code are available for free download from the site `github.com`.

All examples in this chapter were tested under macOS running Julia in version 1.3.1.

8.1 Basics

Launching the Julia application opens a Terminal window, where commands can be entered at the '`julia>`' prompt, very similar to Python.

Example 8.1 As a first example, here is the classical Hello World greeting:

```
julia> println("Hello World")   # my first Julia program
Hello World
```

In Julia, strings are enclosed in (double) quotation marks. The '#' character serves as comment marker just as in Python. A comment block is enclosed in a pair '#=, =#'.

Note that, like in the Python command line interpreter, we can omit the command `println`.

© Springer Nature Switzerland AG 2020
E. Smith, *Introduction to the Tools of Scientific Computing*, Texts in Computational
Science and Engineering 25, https://doi.org/10.1007/978-3-030-60808-8_8

The result of the last input is stored in the special variable ans.

Output is suppressed with a semicolon, e.g. '42;'. The semicolon also serves to separate multiple commands on the same input line.

The history of previous commands can be accessed with the up/down arrow keys.

A program can also be stored in a file, usually with the extension '.jl', and then executed in the Julia shell with, for example, include("hello.jl").

Remark Julia can also be launched directly from the terminal. To do this, we need to make the access path available. On the Mac, this can be done by running the Terminal command

```
$ ln -s /Applications/Julia-1.3.app/Contents/Resources/julia/bin/julia \
      /usr/local/bin/julia
```

Then the application can be launched by typing

```
$ julia
```

in the Terminal command line.

Arithmetical Expressions

Julia is a mathematics oriented language. It provides data types for integers and floating point numbers, by default represented in machine precision, i.e. normally with 64 bits. The type of an object can be inspected with typeof, such as

```
typeof(3)    # ans: Int64
typeof(3.2)  # ans: Float64
```

Note that, similar to the convention in Python, we often write commands without input prompt, and put the answer on the same line, separated by a comment symbol.

As in Python, the result of mixed expressions consisting of integers and floats is always a float. Division of two integers using '/' always returns a floating-point number. There is also a "division from left" (or "reverse division") denoted by '\'.

Integer division is denoted by div, the remainder by '%'. The power operator is '^'. Infix operators can also be written in functional form: '+(1,2)' is equivalent to '1+2'. This is particularly useful in expressions where associative operations can be chained.

Some examples:

```
1 + 2.0     # ans:  3.0
35/5        #       7.0
5\35        #       7.0
2^10        #       1024
div(13, 3)  #       4
13 % 3      #       1
+(1, 2, 3)  #       6
```

As a convenience, a scalar multiplicand for a variable can be written without '*' operator:

```
x = 2; 10x + 4x - 3x/2 + 1  # ans: 26.0
```

Large numbers can also be input in a more readable form, which however is not shown in the response:

```
1_000_007 - 7  # ans:  1000000
```

Variables and Assignments

In the basic form, the assignment to variables works exactly as in Python. Variables are not declared before assigning to them. A variable by itself has no static type, only the value currently stored in it:

```
x = 42;   typeof(x)  # ans: Int64
x = 3.14; typeof(x)  # ans: Float64
```

The standard arithmetic operators also have special updating versions, which can be used to update variables quickly, such as x += 1.

Rational and Complex Numbers

Julia also knows rational and complex numbers.
     Rationals are constructed and represented using the '//' operator:

```
4//3 + 1//2    # ans: 11//6
```

Complex numbers use the constant im to represent the complex number i:

```
1.0im * 1.0im  # ans: -1.0 + 0.0im
```

Mathematical Functions and Constants

Julia is designed as a language for mathematical applications. Therefore, many standard mathematical functions are already part of the core language, such as the square root sqrt, the exponential exp, the absolute value abs, various forms of the logarithmic function, such as the natural logarithm log, and a large collection of trigonometric functions:

```
sqrt(2)  # ans: 1.4142135623730951
exp(1)   #       2.718281828459045
log(2)   #       0.6931471805599453
sin(pi)  #       1.2246467991473532e-16
```

As seen in the last line, the constant π can be accessed by pi. The Euler number e is buried somewhat deeper in the archives as Base.MathConstants.e.

Boolean Operators, Comparisons

The Boolean constants are denoted by true and false, the Boolean operators written as in C: '!' for *negation*, '&&' for *and*, and '||' for *or*.

Comparisons are denoted as in Python. They can be be chained, such that for instance

```
1 < 2 < 3   # ans: true
1 < 2 >= 3  # ans: false
```

8.2 Control Structures: Branching, Loops

Conditional statements and loops are similar to the ones in Python.

Conditional Statements

The basic form of a conditional statement in Julia is

```
if <condition> <statements> else  <statements> end
```

The else branch can again be omitted. Additionally, the basic form may also include one or more elseif clauses. Indentation may be used for readability.

Below we write a Julia implementation of the Collatz problem, discussed in Example 3.3 in the basic Python chapter.

For this we need the distinction "If n is even, then divide it by 2, otherwise multiply n by 3, and add 1".

In Julia this can be formulated by

```
if (n % 2 == 0) n = div(n, 2) else n = 3n + 1 end
```

Note that it is not necessary to enclose the condition in parentheses. It only serves to improve readability.

In short expressions as above, where a conditional choice between single expression values is required, we can use the ternary operator '?:' as in C:

```
n % 2 == 0 ?  n = div(n, 2) : n =  3n + 1
```

The ternary '?:' operator can also be used to assign new values to a variable in an if-else fashion:

```
n = (n % 2 == 0) ? div(n, 2) : 3n + 1
```

This is the form we use below.

While Loop

When we want a loop to run as long as a condition stays true, we use the while loop. The general form is 'while <condition> <statements> end'.

Example 8.2 As an example, here is the Collatz problem:

```
1 n = 100
2 while (n > 1) global n = (n % 2 == 0) ? div(n, 2) : 3n + 1 end
3 println("reached 1")
```

Remark Note the somewhat unintuitive use of the global keyword in line 2. In Julia, commands in code blocks can in general only change local variables that are defined within the block. In interactive programming, the while command starts a new block, so *n* must be explicitly declared as the global variable defined in line 1.

In fact, this applies to while loops and also to for loops discussed below, but not to if-else statements.

For Loop

The general form of a for loop is 'for <e> in <coll> <statements> end', where <e> is an expression, and <coll> can be an array, or any iterable collection to be discussed later.

Example 8.3 Here is the Julia version of our rice-grain computation, discussed in Examples 2.4 and 6.3 in previous chapters:

```
1 sum = UInt64(0)        # max val stored 2^64-1 exceeds  max for Int64
2 fieldval = UInt64(0)   # similarly for max value 2^63
3 for fieldno in 1:64    # range 1 ... 64 incl lower and upper bounds
4     global fieldval = fieldno == 1 ? 1 : fieldval *= 2
5     global sum += fieldval
6 end
7 println(sum)
```

In lines 1 and 2 we define the variables to be of type UInt64, which can store unsigned integers up to $2^{64} - 1$.

The variables sum and fieldval are defined outside of the for loop. They have to be made accessible inside the loop with the global specifier.

Interlude: Ranges

The expression '1:64' is a simple example of a *range*, often used in Julia. Unlike Python, a range includes the upper bound. The general syntax is: '<start>:<end>', or including an increment specifier '<start>:<increment>:<end>'.

When iterating over a numeric range, often '=' is used instead of 'in', such that the counter condition in line 3 can also be written as 'for fieldno=1:64'.

Julia also provides a convenient range-constructing function range, which can take several forms:

```
range(0, stop=1, length = 5)  # ans: 0.0: 0.25: 1.0
range(2, stop=-2, length=9)   #       2.0: -0.5: -2.0
range(1, step=5, length=10)   #       1: 5: 46
range(0, stop=1, step=0.3)    #       0.0: 0.3: 0.9
```

Example 8.4 As a further example of a for loop, we recast the Monte Carlo method for the computation of π in Sect. 4.9 in the SciPy chapter:

```
1  const samples = 10_000_000;  # input
2  hits = 0;
3  for _ = 1 : samples  #  no counter variable  used in loop
4       x, y  = rand(), rand();
5       d = x*x + y*y;
6       if d <= 1 global hits += 1 end;
7  end;
```

In line 1 we declare the 'samples' variable as a constant, whose value will not be changed during computation. This is not necessary, but may help to improve performance.

In line 4 the built-in rand function generates a pair of random numbers from the interval $[0, 1)$. Technically, the comma separated entries form a tuple as in Python. We return to the details later.

The approximation will then result in a value like this:

```
8  println(4*(hits/samples))  # ans: 3.1418
```

The break and continue Statements

As in many other languages, a break statement terminates the loop containing it. Program control flows to the statement immediately after the loop body.

If the break statement is inside a nested loop (loop inside another loop), break will terminate the innermost loop.

The continue statement is used to skip the rest of the code inside a loop for the current iteration only.

8.3 Functions

Functions are the basic building blocks of Julia. A function definition begins with the keyword function, followed by the name of the function and the argument list. What the function does, is defined in the function *body*, a code block terminated by the keyword end. For value types such as numbers, only the argument value is passed to the function. As we shall see later, the behavior is different for reference types such as arrays.

Example 8.5 As a simple example, here is the definition of the factorial function:

```
function factorial(n)  # computes the factorial
    res = 1
    for i = 1:n res *= i end
    return res end;
```

Once defined, the function can be called as usual:

```
factorial(5)  # ans: 120
```

Example 8.6 Functions are not limited to returning a single value:

```
1 function swap(n,m) return (m,n) end
2 swap(1,2)  # ans: (2, 1)
```

As in Python, the expression '(m,n)' in line 1 denotes a *tuple*. We return to tuples in the section on collection types below.

For short functions, there is also a more compact one-line syntax:

```
f(x,y) = x^3 - 2x + x*y; f(3,2)  # ans: 27
```

Anonymous functions are defined as follows:

```
1 x -> x^2  # expression defining anonymous function
2 ans(2)     # ans 4
```

The value of the expression in line 1 is the anonymous function itself. It is automatically stored in the temporary variable ans and can then be used in line 2 to return the result 4.

In the definition of anonymous functions, a single argument can also be enclosed in parentheses, the expression in line 1, for instance, be written as '(x) -> x^2'.

As in Python, an anonymous function can also be assigned to a variable:

```
g = (x,y) -> x^3 - 2x + x*y
g(3, 2)  # ans: 27
```

Recursive Functions

Recursive functions can be defined as in Python.

Example 8.7 As an example, here is the Ackermann function:

```
function ack(x,y)
    if x == 0 return y + 1 end
    if y == 0 return ack(x-1, 1) end
    return ack(x-1, ack(x, y-1))
end
ack(3, 13)  # ans: 131069
ack(4, 1)   # ans: 65533
```

In simple cases, the ternary '?:' operator comes in handy:

Example 8.8 Here is the Fibonacci function:

```
fib(n) = (n == 0 || n == 1) ? 1 : fib(n-1) + fib(n-2);
fib(5)  # ans: 8
```

Example 8.9 In the C++ chapter, we saw that interdependence between function calls had to be resolved using *forward declaration*. In Julia this happens automatically:

```
f(n) = n < 0 ? 22 : g(n);
g(n) = f(n-1);
f(4)  # ans: 22
```

Functions as Arguments and Return Values

As in Python, a function can take another function as argument:

Example 8.10 The following function ddx computes an approximation to the derivative of a function f at a point x:

```
1 function ddx(f, x)
2     h = 1.e-14
3     return (f(x+h) - f(x)) / h
4 end;
5 g(x) = 4x*(1 - x);   # test function
6 ddx(g, 0.5)          # ans: 0.0
```

Recall that in Python a function could again return a function as output value. This is just as easy in Julia:

Example 8.11 Continuing with the program in the last example, we write a function that returns the approximate derivative *function f'* of an input function f:

```
7  function ddx(f)
8      h = 1.e-14
9      function f_prime(x) return (f(x+h) - f(x)) / h end
10     return f_prime
11 end;
```

In line 9 we define a local function f_prime within the body of ddx, and return that function in line 10.

We apply ddx to the same test function g as above and get:

```
12 ddx(g)(0.5)          # ans: 0.0
13 g_prime = ddx(g);
14 g_prime(0.5)         # ans: 0.0
```

Remark Note that in the program above we use the same name 'ddx' for two different functions in lines 1 and 7. This is a simple example of Julia's *multiple dispatch* or *multimethod* feature. In fact, Julia reports ddx to comprise *two* methods:

```
15 ddx   # ans: ddx (generic function with 2 methods)
```

Which of the two methods is to be used depends in the present case on the number of arguments.

Although apparently a simple concept, multiple dispatch depending on the types of values is perhaps the single most powerful and central feature of the Julia language.

Core operations in Julia typically have dozens of methods. As an example we consider the addition operator '+':

```
julia> +
+ (generic function with 163 methods)
```

A complete method list can be obtained with

```
julia> methods(+)
```

Remark Julia is very good at inferring the type of a variable from the context in which it occurs. For instance in lines 3 and 9 in the program above, the argument f is automatically assumed to be a *function* variable.

However, for better performance, it can be beneficial to explicitly specify the variables involved. We can for example declare the argument types in the header in line 1 as

```
function ddx(f::Function, x::Float64)
```

and similarly in line 7:

```
function ddx(f::Function)
```

Interlude: The REPL and Just in Time Compilation

At first sight, the Julia programming interface looks very much like the interpreter environment in Python. It is however a bit more sophisticated. The input is handed over to a so-called *Just in Time*-compiler (JIT), providing an on-the-fly translation to lower level code, which is then immediately executed.

A programming environment such as the one in Julia, is often characterized as a *Read Evaluate Print Loop* (REPL).

Note that using JIT compilation means that the first evaluation of a function will take longer than the following.

Measuring Time, Macros

Execution time can be measured with the built in Julia macro @elapsed, here for example for the factorial:

```
julia> f(n) = (n == 0) ? 1 : n*f(n-1);
julia> @elapsed f(5)   # first run
0.00395807
julia> @elapsed f(5)   # second run
3.131e-6
```

Remark In Julia, macros enable sophisticated code generation that can then be passed to the compiler.

To illustrate this, we give a simplified definition of the @elapsed macro:

```
1 macro myelapsed(expr)
2     return quote
3         t0 = time()
4         $expr
5         t1 = time()
6         t1 - t0
7     end
8 end
9 @myelapsed sin(pi)  # ans: 2.1457672119140625e-6
```

In line 9 the macro @myelapsed is called for the expression 'sin(pi)'. Note that the call requires the preceding '@' (at-sign) before the name of the macro. The effect is that '$expr' in line 4 is replaced by the expression 'sin(pi)', and the resulting code between 'return quote' and 'end' then handed over to the compiler.

Often used is also the extended @time macro, which additionally returns the evaluated expression itself, as well as some more information about resource use.

8.4 Collection Types

In Julia, collections of individual items can be stored in arrays, tuples, dictionaries and sets.

Arrays

Essentially, Julia arrays combine Python lists and NumPy arrays. For now, we'll mainly deal with the list aspects, and later we'll turn to the mathematical properties.

Arrays can be generated as a comma-separated sequence of elements in square brackets:s a comma separated sequence of items enclosed in square brackets:

```
julia> a = [1, 2]
2-element Array{Int64,1}:
 1
 2
```

Here 'a' is declared as a 1-dimensional collection of type Array, consisting of entries of type Int64. The output shows that the array is internally conceived as a *column* vector. In the following we assume that arrays are output with the command print, e.g. print(a), which then returns a in the horizontal form [1, 2].

Note that all entries must be of the same type. If they are not, Julia automatically chooses a lowest common supertype, within a predefined hierarchy of types.

Examples:

```
1 [1, 2.0]      #  is of type Array{Float64,1}
2 [1, 1.0im]    #              Array{Complex{Float64},1}
3 [1, "hello"]  #              Array{Any,1}
```

Remark The type Any in line 3 is a so-called *abstract* data type, there are no objects of type Any. Abstract types only serve as nodes in the type graph. Any is the root of Julia's type hierarchy.

In contrast, *concrete* types such as Float64, String etc. denote types of actual objects.

Array entries are accessed in bracket notation a[n] as in Python, where however indices begin with 1 according to usual mathematical conventions.

The number of items in an array a can be inquired by length(a), the last element by a[end]:

```
a = [3, 2, 1];
length(a)   # ans: 3
a[end]      # ans: 1
```

As in Python, Array is a *reference* type. A *copy* b of a is obtained with b = copy(a).

Arrays can also be created directly from range objects with the function collect:

```
1  a = collect(3: 5)          # ans:  [3, 4, 5]
2  b = collect(3: 2: 8)       # ans:  [3, 5, 7]
```

Note that, unlike Python, in line 1 both bounds are included.

For *float* ranges an increment value of 1.0 is assumed by default:

```
3  c = collect(0.0 : 1.5)     # ans:  [0.0, 1.0]
4  d = collect(1.0: -0.3 :0)  # ans:  [1.0, 0.7, 0.4, 0.1]
```

Arrays can be concatenated using the function vcat, where the initial 'v' for 'vertical' reflects that arrays by default are conceived as column vectors.

Example:

```
a = [1, 2, 3]; b = [4, 5];
b = vcat(a,b)  # ans: [1, 2, 3, 4, 5]
```

An empty array is denoted by [ ]. Note that in this generic form the array will be of type Any. If we want an empty array of a certain type, such as integer, we need to specify it like this:

```
a = Int64[]    # 0-element Array{Int64,1}
```

or equivalently simply as a = Int[], since on 64-bit machines the integer type defaults to Int64.

Arrays can hold entries of arbitrary types. Very often, arrays of arrays are needed, such as

```
a = [[1, 2], [3, 4], [5, 6]];
typeof(a)        # ans: Array{Array{Int64,1},1}
```

Entries can then for instance be accessed by double bracket notation:

```
print(a[end])  # ans: [5, 6]
a[end][1]      # ans: 5
```

Arrays and Mutating Functions

As mentioned, Array is a reference type. In Julia, objects of reference types are passed to a function not by value but by reference. Any change made within the function is applied globally, the function *mutates* the object.

Example 8.12 Here is a mutating function that swaps two entries in an array:

```
function swap!(a) a[1], a[2] = a[2], a[1] end
arr = [1, 2];
swap!(arr); println(arr)  # ans: [2, 1]
```

Note that the function swap! has no return value. It accesses the external variable arr and performs the swap *in-place*.

Also note that functions that change one or more of their arguments are usually given an exclamation mark as part of the name.

Here is a built in Julia function 'push!' that we will need later. It appends a new entry to the end of the array:

```
a = [1, 2];
push!(a, 42);
println(a)  # ans: [1, 2, 42]
```

Filtering

Below we use arrays to formulate a Julia program to compute the sieve of Eratosthenes, discussed in the Python chapter. For this we can use the powerful filter function for arrays.

If a is an array and f a Boolean function, then the array a filtered by f results from a by removing all elements x for which f(x) is false.

Actually, there are two variants of the filter function: filter, which returns a *new* array, and a mutating function filter! (again written with an exclamation mark), which *modifies* the argument array.

Filtering is often used with anonymous Boolean functions of the form 'x -> <Bool expression>'.

Example 8.13 Filtering with the function 'x -> x % 2 == 0' returns the even numbers in the argument array:

```
a = collect(1: 5)
filter(x -> x % 2 == 0, a)   # ans: [2, 4]
print(a)                     # ans: [1, 2, 3, 4, 5], 'a' not modified
filter!(x -> x % 2 == 0, a)
print(a)                     # ans: [2, 4], 'a' modified in-place
```

Example 8.14 Here is a Julia implementation of the sieve of Eratosthenes from Example 3.4 in the Python chapter:

```
1 n = 30
2 L = collect(2:n)
3 P = Int[]        #  empty integer list
4 while L != []
5     p = L[1]     # the smallest number still contained in L
6     push!(P, p)  # is appended to the end of P
7     for i in L   # removes all multiples of p
```

```
8              if i % p == 0  filter!(x -> x != i, L) end
9       end
10  end
11  println(P)
```

Note that we initialize P in line 2 as an empty *integer* array. The program would work also with the initial assignment 'P = []'. Then, however, the type Any would prevail throughout the whole program.

Remark The attentive reader may have noticed that here, in contrast to the loop constructions before, we have used the variables L and P inside the loop body without declaring them as global. The deeper reason is again that Array is a reference type, where references to memory addresses are passed, not values.

List Comprehension

The comprehension methods for lists discussed in Python also exist in Julia.

Example 8.15 Here is a Julia version of the Quicksort algorithm:

```
1  function qsort(lst::Array{Int})
2      if lst == [] return Int[] end
3      p = lst[1]
4      sml = qsort([x for x in lst[2: end] if x < p])
5      grt = qsort([x for x in lst[2: end] if x >= p])
6      return vcat(sml, [p], grt)
7  end;
```

A test run:

```
8  testList = [4, 5, 7, 3, 8, 3];
9  println(qsort(testList))
```

In line 2 the base case in the recursion again refers to an empty *integer* array. Otherwise, the type 'Any' would propagate into the result.

The type specification '::Array{Int}' for the argument, on the other hand, is not strictly necessary. It only serves as a hint to the compiler to optimize the code.

Tuples

Like arrays, a tuple is an ordered sequence of elements, with the difference that it cannot be changed after it has been created. A tuple is represented by a comma-separated sequence of items enclosed in parentheses, where the parentheses can actually also be left out. As in Python, a tuple of length 1 must be distinguished from its content by a trailing comma.

Examples:

```
a = (2.0, "hello");   typeof(a)  # ans: Tuple{Float64,String}
b = 1, 2;             typeof(b)  #      Tuple{Int64,Int64}
c = (1);              typeof(c)  #      Int64
d = (1,);             typeof(d)  #      Tuple{Int64}
```

Dictionaries

Dictionaries essentially work like in Python, except that the separator ':' is replaced by '=>':

```
dict = Dict("a" => 1, "b" => 2, "c" => 3)
```

Recall, however, that strings in Julia are always included in double quotes.

Sets

Sets can be created from arrays:

```
a = Set([3, 3, 1, 4])  # ans: Set([4, 3, 1])
```

The empty set is defined by

```
Set()  # ans: Set(Any[])
```

or specifically, as an empty set of, say, integers as

```
Set{Int}()  # ans: Set(Int64[])
```

Test for membership is performed with 'in':

```
a = Set([3,3,4,1]);
1 in a  # ans: true
```

The standard set operations union, intersection, and difference, are performed with the functions union, intersect, and setdiff. Single elements can be added with 'push!'.

8.5 Composite Types

Unlike Python and C++, Julia does not support object oriented programming. As in C, data types can be created with *structs*, but they do not allow access methods to be bundled in encapsulated functional units as in classes. Instead, Julia relies on its key concept *multiple dispatch* to specify type dependent behavior of functions.

Julia Structures

Julia structures resemble the ones in C, where however now the entry types do not have to be specified.

Example 8.16 (Linked Lists) We illustrate the idea with an example already discussed in the C and C++ chapters, the construction of linked lists for the storage of sparse vectors.

Consider the vector

```
1 v = [0, 2, 0, 0, 8, 10, 0, 14, 16, 18];
```

We define a type that can store pairs of indices and values:

```
2 struct node
3     index::Int
4     value::Int
5 end
```

In general, in Julia the definition of a struct type consists of a list of components together with their type, where again the type Any is assumed, if no type is specified.

Remark Note that the type is here 'node' and not 'struct node' as in C, where we had to apply the typedef operator to get the abbreviated notion.

The sparse representation is now defined as a list of all index-value pairs for indices with non-zero values.

We begin with an empty list, prepared to store entries of type node:

```
6 sp = node[];
```

The list is then successively populated with nodes that have non-zero values:

```
7 for i in 1 : length(v)
8     if v[i] != 0 push!(sp, node(i, v[i])) end end
```

In the loop each index-value pair with a non-zero value is stored as a new node and appended to the list sp.

To test the construction we print the resulting representation:

```
9 for nd in sp print("($(nd.index), $(nd.value)) ") end
```

and obtain

```
10 (2, 2) (5, 8) (6, 10) (8, 14) (9, 16) (10, 18)
```

Interlude: String Interpolation

In line 9 above we use *string interpolation* to insert the node components into the string to be printed. In general, any Julia expression can be inserted into a string with the '$()' construction, where it is then evaluated and replaced by the result. The parentheses can even be omitted for individual variables.

Example:

```
a = 42;
print("The answer is $a")   # ans: The answer is 42
```

Note that we have actually encountered string interpolation before, in the definition of the macro myelapsed at the end of Sect. 8.3.

Mutable Structures

Julia structures are immutable, they cannot be changed, once created. However, Julia also provides a *mutable variant*.

Example:

```
mutable struct point
    x::Int; y::Int
end
p = point(1, 2);
p.x = 4
print(p)  # ans: point(4, 2)
```

Structures vs. Classes

As mentioned, Julia does not support object oriented programming. In the following example, we illustrate how to emulate certain class constructions using structs.

Example 8.17 (Fractions) Consider the definition of the class Fraction in Example 3.18 in the Python chapter. We show how a to create a corresponding data type in Julia.

The data structure itself is defined as in Python, now however based on a struct type:

```
1 struct fract num::Int; den::Int end
```

Variables can then be initialized and declared for instance as

```
2 a = fract(1, 2); b = fract(3, 4);
```

In a class definition, we could now add object specific methods to the data structure. This is not possible in Julia. We have to define the methods *globally*, however in such a way that they apply only to fract objects. This is where the multiple-dispatch approach comes to help.

First we define a function add to compute the sum of two fractions:

```
3 add(x::fract, y::fract) = fract(x.num*y.den+x.den*y.num, x.den*y.den)
```

Note that the function is now exclusively applicable to arguments of type fract.

We test it:

```
4 c = add(a,b)  # ans: fract(10, 8)
```

In Python we could use operator-overloading, so we could write the addition in the form 'a + b'. In Julia we can achieve this through the multiple-dispatch feature.

Recall that the '+' operator is a generic function with 163 methods, where the choice of the actual method depends on the number and type of arguments.

Now we can actually *add* another method to '+', which applies specifically to fractions. This is achieved as follows:

```
5 import Base: +
6 +(x::fract, y::fract) = fract(x.num*y.den + x.den*y.num, x.den*y.den)
7 +  # ans: + (generic function with 164 methods)
8 a + b  # ans: fract(10, 8)
```

In line 5 we import the current definition for the '+' function. The general specification declares '+' to be applicable as infix operator.

In line 6 we add a new specific method that is to be applied whenever the arguments are of type fract. The output in line 7 shows that the number of methods in the '+' function has now been increased by one.

In fact, now the add definition in line 3 is no longer needed, and can be deleted.

To test fractions for equality we can proceed similarly:

```
 9  import Base: ==
10  ==(x::fract, y::fract) = x.num*y.den == x.den*y.num
11  a == b                         # ans: false
12  fract(1, 2) == fract(2 ,4)     # ans: true
```

Remark In the Python chapter, we also discussed a class Polynomial with a *subclass* Parabola. To recast the concept of inheritance is however not immediate in Julia.

8.6 Linear Algebra

Julia is designed specifically to address the needs of high-performance numerical analysis and computational science. So it comes as no surprise that already the core language offers a large collection of linear algebraic tools.

Similar to NumPy, vectors and matrices are based on the Array type.

Recall that an array a = [1,2,3] is conceived as a column vector. In contrast, an array defined without separating commas is interpreted as row vector, more precisely as a 1-row array of dimension 2, i.e. a matrix:

```
julia> [1 2 3]
1×3 Array{Int64,2}:
 1  2  3
```

Also in general, matrices can be generated via the bracket notation. More precisely, a matrix can be defined as a sequence of row vectors of equal length, separated by a semicolon:

```
julia> A = [1 2 3; 4 5 6; 7 8 9]
3×3 Array{Int64,2}:
 1  2  3
 4  5  6
 7  8  9
```

Matrices can also be constructed using comprehension, for instance the 3×3 Hilbert matrix with rational number entries like this:

```
julia> H3 = [1//(i+j-1) for i = 1:3, j = 1:3]
3×3 Array{Rational{Int64},2}:
 1//1  1//2  1//3
 1//2  1//3  1//4
 1//3  1//4  1//5
```

Here is a list of common matrix operations. Recall that in Julia indices begin at 1:

```
A = [1 2; 3 4]
A[2,1]    # entry value 3, recall: indices start at 1
A'        # transpose of A, also written as transpose(A)
2 * A     # scalar multiplication
B = [5 6; 7 8]
A + B     # addition
A * B     # matrix multiplication
A .* B    # componentwise multiplication
```

Note that the componentwise multiplication operator is written as '.*'. Also note that adding component by component with a scalar in the form, e.g. 2 + A, is not supported. For the componentwise addition, the explicit component operator '.+' is required.

If f is a function defined for single arguments and A is a matrix, then f.(A) is the matrix that results from componentwise application of f to the entries of A. This is also known as *broadcasting*.

Julia also offers *matrix division* from left 'B\A' and right 'A/B'. According to the documentation, this corresponds to the computation of $B^{-1}A$ and AB^{-1}.

Again, if the operations are to be performed componentwise, the operators must be preceded with a dot as '.\' or './'.

Linear Equations

Division from left can be used to solve linear equations:

```
A = [9. 3. -6. 12.; 3. 26. -7. -11.; -6. -7. 9. 7.; 12. -11. 7. 65.];
b = [ 18.;   11.;   3.;   73.];
println(A \ b)   # ans: [1.0, 1.0, 1.0, 1.0]
```

Conjugate Gradient

Note that the matrix A above is symmetric and positive definite. We use this as an opportunity to solve the equation by the conjugate gradient method, discussed in Example 4.2 in the SciPy chapter.

Example 8.18 (Conjugate Gradient) Here is the Julia program:

```
# include A and b from above
n = length(b)
x = zeros(n)
p = r = b
rs_old = r'*r
for _ in 1:n
    Ap = A*p
    alpha = rs_old / (p'*Ap)
    global x += alpha*p
    global r -= alpha*Ap
```

```
      rs_new = r'*r
      if sqrt(rs_new) < 1e-10 break end
      global p = r + (rs_new / rs_old)*p
      global rs_old = rs_new
end
println(x)  # ans: [1.0, 1.0, 1.0, 1.0]
```

Again note the 'global' specifiers, needed to provide access to the external variables from within the loop.

Special Matrices

Matrices consisting only of 0 or 1-entries can be constructed by zeros(n,m) and ones(n,m).

As in NumPy, 0- and 1-*vectors* are constructed by zeros(n) and ones(n).

There is no special notion for the identity matrix. It can be constructed using the Diagonal function introduced below.

Submatrices

To extract submatrices, we can use the method of *slicing*:

```
A = [1 2 3 4; 5 6 7 8; 9 10 11 12; 13 14 15 16];
A[3:4, 2:3]  # ans:
    2×2 Array{Int64,2}:
    10  11
    14  15
A[1, :]                 # returns 1st row
A[1:3, :]               # returns rows 1 through 3
A[:, end-1 : end]       # returns last 2 columns
i = [1, 3]; A[:, i]     # returns columns 1 and 3
```

Reshape

As in NumPy, we can rearrange the entries in a matrix, here with the function 'reshape':

```
A = [1 2 3 4 5 6 7 8; 9 10 11 12 13 14 15 16];
reshape(A, (2, 8))
ans: 1  3  5  7   9  11  13  15
     2  4  6  8  10  12  14  16
```

The result corresponds to what we got with the order='F' option in NumPy.

To flatten a matrix to a linear vector we can also use the 'vec' function:

```
vec(A)  # result represented horizontally:
1  9  2  10  3  11  4  12  5  13  6  14  7  15  8  16
```

As usual in Julia, the result is actually a column vector, in order to save printing space here displayed horizontally. Note again the column-major element order.

Application of vec(A') returns the elements in a row-major order 1 2 3 4 5 6 7 8 9 10 11 12 13 14 15 16.

The Standard Library Module LinearAlgebra

So far we have limited ourselves to the built-in linear algebraic tools. For more advanced constructions we have to use the LinearAlgebra package which is included in the standard Julia distribution. It is loaded with

```
using LinearAlgebra
```

The LinearAlgebra module offers array arithmetic, matrix factorizations and other linear algebra related functionality.

Now we can for instance compute the determinant of a matrix with det(A) and the inverse with inv(A). Eigenvalues and -vectors can be found with eigvals(A) and eigvecs(A). With the help of the Boolean functions issymmetric(A) and isposdef(A) we can test if a matrix is symmetric or positive definite.

Standard factorization methods are provided by lu(A) for LU-, cholesky(A) for Cholesky-, and qr(A) for QR-decomposition.

The Poisson Matrix

In the SciPy chapter we presented the Poisson matrix, which is used to implement the finite difference method for solving partial differential equations. For details we refer to the discussion at the end of Sect. 4.1. In the following we present a corresponding Julia construction.

Example 8.19 (Poisson Matrix) We define a function poisson that for a given number n constructs the $n^2 \times n^2$ Poisson matrix:

```
 1  using LinearAlgebra
 2  function poisson(n::Int)
 3      v = 4*ones(n);
 4      w = ones(n-1);
 5      B = SymTridiagonal(v, -w);
 6      sB = SymTridiagonal(zeros(n), w);
 7      Id = Diagonal(ones(n));
 8      A = kron(Id, B) + kron(sB, -Id);
 9      return A
10  end
```

Line 3 creates a vector v of length n, which consists of constant 4-entries. Line 4 results in a 1-vector of length $n - 1$. In line 5 we use the function SymTridiagonal to define a symmetric tridiagonal matrix B with 4s in the diagonal and -1s in both the first super- and subdiagonal.

In line 6 we construct a symmetric tridiagonal matrix sB with 0-entries in the diagonal and 1-entries in the first super- and subdiagonal.

Line 7 defines the $n \times n$ identity matrix. In general, for a vector x, `Diagonal(x)` defines a matrix with the x-entries in the diagonal and 0-entries elsewhere.

In line 8 the Kronecker product is applied twice and the results added up to yield the desired Poisson matrix.

The construction can then be tested by

```
11  A = poisson(3)
```

This gives exactly the same result as in the SciPy chapter.

Sparse Matrices

As mentioned earlier, the use of sparse matrices and matrix operations can result in significant performance gains. Below, we consider a sparse version of the Poisson matrix that is used to solve Poisson equations using the finite difference method.

To get a sparse version of the Poisson matrix, we could start from scratch and use sparse variants of the construction methods. Most of the performance gain, however, comes from optimizing the process of solving the linear equation itself. We therefore choose to construct our matrix as a full matrix and then convert it to a sparse representation.

Example 8.20 (Sparse Poisson Matrix) We continue with the previous example:

```
12  using SparseArrays  # Loading module
13  spA = sparse(A)      # converting poisson matrix in line 11
```

In line 12 the standard library module `SparseArrays` is loaded, which in particular contains the conversion function `sparse`, used to generate the sparse variant spA of the Poisson matrix in line 13.

The output is

```
14  9×9 SparseMatrixCSC{Float64,Int64} with 33 stored entries
```

where the entries are of the form

```
15  [1, 1] = 4.0, [2, 1] = -1.0 , ... , [8, 9] = -1.0, [9, 9] = 4.0
```

Line 14 explains that the internal representation is in *compressed sparse column* format CSC. Line 15 shows that the entries are the desired ones.

Example 8.21 We continue with the program in the last example and compare the performance of full and sparse Poisson matrices in solving a simple system of linear equations. More precisely, for $n = 100$ we solve $Ax = b$ for the $n^2 \times n^2$ Poisson matrix A and the 1-vector b of length n^2. Time is again measured with the Julia-macro `@elapsed`:

```
16 n = 100;
17 A = poisson(100);
18 b = ones(n*n);
19 @elapsed A \ b      # ans: 3.548505505
20 spA = sparse(A);
21 @elapsed spA \ b  # ans: 0.021005501
```

This corresponds to a speed gain factor of approximately 170.

8.7 Ordinary Differential Equations

Julia has no built-in tools for solving differential equations, so we have to install an additional package. Here we use the package `DifferentialEquations`, developed by Chris Rackauckas, at the University of California-Irvine. Extensive documentation can be found at `juliadiffeq.org`.

We need to download and install the package. This must be done only once.

```
1 import Pkg;                       # Julia's package manager
2 Pkg.add("DifferentialEquations")  # only required once
```

In line 1 the internal package manager is loaded. In line 2 the `Pkg`-function 'add' downloads and installs the external package.

The commands 'import' and 'using' are closely related: 'import ...' behaves just as it would in Python, whereas 'using ...' corresponds to 'from ... import *' in Python.

Once the package installation is complete, we can use it like any built-in module with

```
3 using DifferentialEquations  # required once in every session
```

Example 8.22 As a first example, we formulate the Julia solution of a simple initial value problem

$$u'(x) = x - u(x), \quad x \in [0, 5],$$
$$u(0) = 1,$$

with the exact solution $u(x) = x - 1 + 2e^{-x}$.

We assume the package `DifferentialEquations` already to be loaded according to lines 1–3 above.

The problem is then encoded by

```
4 dudx(u, p, x) = x - u  # note additional parameter p
5 bpts = (0.0, 5.0)       # interval given by tuple of boundary points
6 u0 = 1.0                # initial value
7 prob = ODEProblem(dudx, u0, bpts)
```

The parameter p in line 4 is required, even if not used. We return to that below. The interval where the equation is to be evaluated, is denoted by a pair of boundary values. The function `ODEProblem` stores the equation in the variable prob.

The problem is then passed to the solver `solve`. The solver can use different approximation algorithms. If none is specified as here, a default method is selected:

```
8 sol = solve(prob)
```

To plot the solution we need another external package. A popular choice is the package `Plots`, which works very well with `DifferentialEquations`. Documentation can be found at the official site `juliaplots.org`.

```
9  Pkg.add("Plots")
10 using Plots
11 plot(sol, leg=false)  # no legend shown
```

For comparison, we also plot the exact solution, computed at the evaluation points selected by the algorithm. Note that we can access the array of evaluation points with `sol.t`. As with the solver `solve_ivp` in SciPy, `DifferentialEquations` assumes differential equations to evolve in a *time span*, represented by a variable t:

```
12 plot!(sol.t, t-> t-1+2*exp(-t), seriestype=:scatter, leg=false)
```

Note again that the exclamation mark in `plot!` indicates that the operation works in-place, i.e. *modifies* the original plot. The result is shown in Fig. 8.1.

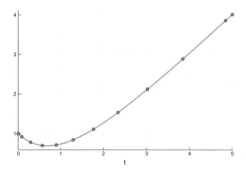

Fig. 8.1 Julia solution of initial value problem in Example 8.22. Solid dots indicate exact solution at the supporting points determined by the approximation algorithm

The plot can then be saved to a file, e.g. in portable document format PDF, with

```
13 savefig("odeplot.pdf")
```

Equation Systems

As an example of a *system* of first order differential equations, we return to the Lotka–Volterra equations from the Scientific Python chapter.

Example 8.23 (Lotka-Volterra) In general, Lotka–Volterra equations are defined by

$$u_1' = \alpha u_1 - \beta u_1 u_2,$$
$$u_2' = -\gamma u_2 + \delta u_1 u_2,$$

where u_1 and u_2 may be assumed to describe the development of a prey-predator population over time.

In the following, we fix the parameters $\alpha = 1.0$, $\beta = 0.2$, $\gamma = 0.5$, $\delta = 0.2$, set the time span to $[0, 4\pi]$, and define the initial values as $u_1(0) = 1.0$, $u_2(0) = 2.0$.

We assume the packages DifferentialEquations and Plots to be loaded.

The parameter values are encoded in a tuple p:

```
1 p = (1.0, 0.2, 0.5, 0.2);
```

Time span and initial values are given as

```
2 tspan = (0.0, 4pi); u0 = [1.0, 2.0];
```

In the previous example, we specified the equation with an explicit definition of the derivative function dudt. When dealing with equation *systems*, dudt will be an *array*. Hence we can take advantage of an implicit in-place specification

```
3 function lotvol!(dudt, u, p, t)
4     dudt[1] =  p[1]*u[1] - p[2]*u[1]*u[2]
5     dudt[2] = -p[3]*u[2] + p[4]*u[1]*u[2]
6 end
```

The derivative specification can then be handed over to ODEProblem like this:

```
7 prob = ODEProblem(lotvol!, u0, tspan, p);
```

The problem is now solved as before:

```
8 sol = solve(prob);
```

The solution can then be passed to the plotter. It will plot both functions u_1 and u_2 (see Fig. 8.2):

```
9 plot(sol)
```

Second Order ODEs

As in Python, we transform second order equations into a system of first order equations.

Example 8.24 We consider a simple form of an oscillation equation (with the obvious solution $u = \sin$):

$$u'' + u = 0 \ \text{ in } [0, 2\pi],$$
$$u(0) = 0, \ u'(0) = 1.$$

If u[1] represents the unknown function u itself and u[2] the derivative u', we can set

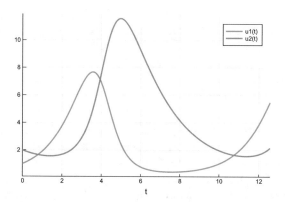

Fig. 8.2 Julia solution of the Lotka-Volterra equation in Example 8.23

```
1 function u!(dudt, u, p, t) dudt[1] = u[2]; dudt[2] = -u[1] end;
2 tspan = (0.0, 2pi); u0 = [0.0, 1.0];
```

The system can then be solved as usual, assuming `DifferentialEquations` to be loaded:

```
3 prob = ODEProblem(u!, u0, tspan);
4 sol = solve(prob);
```

It can then be plotted. Note that `plot(sol)` will show the solutions for both u and u'.

Boundary Value Problems BVPs

In Julia, too, boundary value problems are more difficult to solve than initial value equations.

As a simple example we return to the equation from Example 4.12 in the Scientific Python chapter:

Example 8.25 Consider the BVP

$$u''(x) = 6x \ \text{ in } [0, 1],$$
$$u(0) = 0, \ u(1) = 1,$$

with the obvious solution $u(x) = x^3$.

The equation is again transformed into a system of first order equations, formulated in in-place form:

```
1 function u!(dudx, u, p, x) dudx[1] = u[2]; dudx[2] = 6x end;
```

The boundary points:

```
2 bds = (0.0, 1.0);
```

The boundary values $u(0) = 0$, $u(1) = 1$ are encoded in a function 'bc!' in *residual* form that describes what has to be subtracted to obtain a homogeneous system. During the approximation, bc! will be called repeatedly:

```
3 function bc!(residual, u, p, x)
4     residual[1] = u[1][1]
5     residual[2] = u[end][1] - 1
6 end;
```

Note that u[1] in line 4 refers to the current value pair at the beginning of the time span, where u[1][1] stores the current approximation to $u(0)$. In the solution process the value of residual[1] should converge to 0.

Dually, u[end] in line 5 refers to the last element in the value-pair array, where u[end][1] stores the current approximation to $u(1)$. The correct value is again reached when residual[2] converges to 0.

To formulate the BVP in Julia we need the function TwoPointBVProblem:

```
7 bvp = TwoPointBVProblem(u!, bc!, [0, 0], bds);
```

The function expects an initial guess for the solution. The pair [0,0] declares the start values to be constant zero over the whole interval for both involved functions.

The solution is again obtained with the multipurpose solver solve. Presently, however, the solver for BVPs can only deal with a particular algorithm MIRK4. Moreover, a user specified mesh has to be supplied:

```
8 sol = solve(bvp, MIRK4(), dt=0.05);   # note: 'dt' with 't' required
```

The solution containing both u and u' can then be plotted as usual.

Example 8.26 (Bratu equation) In Example 4.14 in the SciPy chapter, we solved the Bratu equation

$$u''(x) = -e^{u(x)}, \quad x \in [0, 1],$$
$$u(0) = u(1) = 0.$$

Recall that the equation has *two* solutions. Here is the Julia version of the solution program:

```
function u!(dudx, u, p, x) dudx[1] = u[2]; dudx[2] = -exp(u[1]) end;
bds = (0.0, 1.0);
function bc!(residual, u, p, x)
    residual[1] = u[1][1]
    residual[2] = u[end][1]
end;
solinit = [0, 0]   # for second solution set solinit = [3, 0]
bvp = TwoPointBVProblem(u!, bc!, solinit, bds);
sol = solve(bvp, MIRK4(), dt=0.05);
plot(sol)
```

Fig. 8.3 shows both solutions, to the left the one based on the initial guess [0,0], to the right the one based on [3,0]. Note that the designator 't' for the argument parameter is not easily changed.

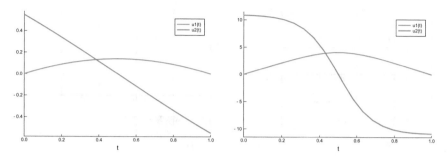

Fig. 8.3 Two solutions for the Bratu equation in Example 8.26. The blue lines show the solution functions, the red ones the derivatives.

8.8 Partial Differential Equations

In Sect. 4.8 in the SciPy chapter we discussed how to solve partial differential equations PDEs with the *finite difference method*, illustrated for *2D Poisson equations* with *Dirichlet boundary conditions*. Here we develop a corresponding Julia program.

Recall that the general form of such a 2D Poisson problem is to determine a function $u: \Omega \to \mathbb{R}$, $\Omega := [0, 1] \times [0, 1]$, such that for some given functions $f, g: \Omega \to \mathbb{R}$:

$$\begin{aligned} -\Delta u &= f \quad \text{in } \Omega, \\ u &= g \quad \text{on the boundary } \partial\Omega. \end{aligned}$$

(1)

For background information we refer to the introduction in Sect. 4.8, and turn directly to the solution program, written as a Julia *function*.

The Solution Program

We define a Julia function `poisson_solver` which for given functions f, g and integer numbers m returns an approximation to (1) on an $m \times m$ grid.

We begin with the preamble:

```
1 using LinearAlgebra, Plots
```

For convenience, we repeat the definition of the function to generate the Poisson matrix from Example 8.19 above:

```
2 function poisson(n::Int)
3     v = 4*ones(n);
4     w = ones(n-1);
5     B = SymTridiagonal(v, -w);
6     sB = SymTridiagonal(zeros(n), w);
```

```
7        Id = Diagonal(ones(n));
8        A = kron(Id, B) + kron(sB, -Id);
9        return A
10  end
```

We come to the solver function:

```
11  function poisson_solver(f::Function, g::Function, m::Int)
```

The meshgrid and the distance between grid points h is given by

```
12       x = y = range(0, stop=1, length=m);
13       h = 1/(m-1);
```

The solution matrix u is initialized with zeros, then modified by the g-values on the boundaries:

```
14       u = zeros(m,m);
15       for i = 1:m, j = 1:m  # nested loops
16           if i ==  1 || i == m || j == 1 || j == m
17               u[i,j] = g(x[i], y[j])
18           end
19       end
```

Note the concise form of the nested loop in line 15.

Next comes the encoding of the source function f:

```
20       n = m-2  # number of inner points
21       F = zeros(n,n)
22       for i = 1:n, j = 1:n F[i,j] = f(x[i+1], y[j+1]) end
```

The boundary adjacent points of F are then modified by the g values stored in the u-boundary:

```
23       F[:,1] += u[2:n+1, 1]    / h^2
24       F[:,n] += u[2:n+1, n+2] / h^2
25       F[1,:] += u[1, 2:n+1]    / h^2
26       F[n,:] += u[n+2, 2:n+1] / h^2
```

In the matrix equation we need F flattened to a vector:

```
27       b = vec(F)  # recall: vec generates column-major order
```

We can now apply the Poisson matrix to solve the equation:

```
28       A = poisson(n);
29       u_inner = A \ b*h*h
```

The solution u_{inner} is then inserted in the solution matrix u and the result is returned as a tuple of x, y and u:

```
30       u[2:n+1, 2:n+1] = reshape(u_inner, n, n);   # column-major order
31       return x, y, u
32  end; # poisson_solver
```

As a test we solve (1) for

(2) $$f(x, y) := 1.25e^{x+\frac{y}{2}}, \quad g(x, y) := e^{x+\frac{y}{2}} :$$

```
33  f(x,y) = 1.25*exp(x + y/2);
34  g(x,y) = exp(x + y/2);
35  sol = poisson_solver(f, g, 17);
36  plot(sol, seriestype=:wireframe)
```

The result is shown in Fig. 8.4.

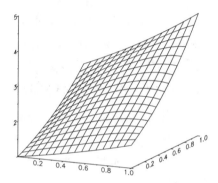

Fig. 8.4 Julia solution of the Poisson equation (1) with source function f and boundary function g in (2)

8.9 Working with Files

File management in Julia is similar to that in Python. Here is a simple example:

```
f = open("test.txt", "w")          # opens file in writing mode
write(f, "A line\n Another one")   # writes string to file
close(f)                           # closes connection to file
f = open("test.txt")               # opens file in reading mode
s = read(f, String)                # assigns file content to string s
```

The main difference to Python is that the functions `read`, `write` and `close` now take the file handler as *argument*, while in Python the class *methods* are appended to the handler *object* in dot-notation.

Julia Set Fractals

As a more interesting example, we develop a program to store a graphical bitmap figure in a normal text file in "portable graymap format" PGM. A detailed explanation of that format can be found, for instance, in the Wikipedia article `Netpbm_format`.

When programming in Julia, it is too tempting to consider Julia set graphics, more precisely *Julia set fractals*, where the name accordance is probably purely coincidental. Julia sets are named after the French mathematician Gaston Julia. Julia set fractals are closely related to the Mandelbrot set, which we will discuss later in the exercises in the MATLAB chapter.

Julia set fractals are normally generated by initializing a complex number $z = z_0 = x + yi$ where x and y are image pixel coordinates in the range of about -2 to 2. Then, z is repeatedly updated using $z_{n+1} := z_n^2 + c$ where c is another complex number that gives a specific Julia set. After numerous iterations, if the magnitude of z is less than 2, we say that the pixel is in the Julia set and color it accordingly. Performing this calculation for a whole grid of pixels gives a fractal image.

Example 8.27 (A Julia Program to Generate Julia Sets) Specifically, we consider the iterating function $f: \mathbb{C} \to \mathbb{C}$ given by

$$f(z) := z^2 + c, \quad c := -0.8 + 0.156i.$$

The Julia program begins with the definition of the iterator constant:

```
1 const c = -0.8 + 0.156im;
```

We open a file `julia.pgm` in writing mode and connect it to a file handler `fout`:

```
2 fout = open("julia.pgm", "w");
```

Next we declare the image to be of both width and height s. The entire image specification is then written to the file preamble:

```
3 s = 1000;
4 write(fout, "P2\n# Julia Set image\n$s $s \n255\n");
```

Basically, the string in line 4 says that each pixel in the $s \times s$ grid will be associated with a grayscale value, where 0 corresponds to black, and 255 to white.

The main computation takes place within the nested block in lines 5–12:

```
5  for y in range(2, stop=-2, length=s)       # Im-rows, top to bottom
6      for x in range(-2, stop=2, length=s)   # Re-columns, left to right
7          z = complex(x,y); n = 0;
8          while (abs(z) < 2 && n < 255) z = z*z + c; n += 1 end
9          write(fout, "$(255-n) ")
10     end
11     write(fout, "\n")
12 end
```

In line 7 each pixel point (x, y) is interpreted as a number z in the complex plane, and the iteration counter set to 0. The while loop in line 8 tests whether the pixel can be verified to *not* belong to the set in less than 256 steps, and if so, records the number of necessary steps in the variable n. In line 9 a corresponding gray value is output to the file, increasingly darker with the number of iterations.

The nested for loops beginning in lines 5 and 6 range over all grid points, in the outer loop from top to bottom for each imaginary part y, in the inner loop then horizontally for each real part x.

Note that in previous Julia versions, we could use a `linspace` construction to define the range. However, the `linspace` function is no longer available.

In line 11 each horizontal grid row is terminated by a line break.

It remains to close the file-connection:

```
13 close(fout)
```

The file content interpreted as a PGM representation is shown in Fig. 8.5.

Fig. 8.5 Julia set generated with iteration function $z \mapsto z^2 - 0.8 + 0.156i$

8.10 Exercises

Monte Carlo Method

Exercise 8.1 Write a Julia function that uses a Monte Carlo method to compute the value of the integral $\int_0^1 f(x)\,dx$ for $f: \mathbb{R} \to \mathbb{R}$ with $f([0,1]) \subseteq [0,1]$. Test the program for $f: x \mapsto e^{-x}$.

Continued Fractions

A (finite) *continued fraction* is an expression of the form

$$a_0 + \cfrac{1}{a_1 + \cfrac{1}{a_2 + \cfrac{1}{\cdots + \cfrac{1}{a_n}}}},$$

where $a_i \in \mathbb{Z}$, $a_1, \ldots, a_n \geq 1$. We denote this fraction by $[a_0, a_1, \ldots, a_n]$.

Exercise 8.2 Write a Julia function to compute the rational number a/b given by a continued-fraction representation $[a_0, a_1, \ldots, a_n]$.

Exercise 8.3 Conversely, write a function that converts a rational number a/b to a continued fraction.

Note that the numbers in the continued fraction are precisely the quotients that would arise, if we used the Euclidean algorithm to compute $\gcd(a, b)$.

Here the Julia function divrem might come in handy, that returns quotient and remainder from Euclidean division.

Exercise 8.4 Write a function that approximates a number $r \in \mathbb{R}$ by a continued fraction of the form $[a_0, a_1, \ldots, a_n]$ for a given n.

Here Julia's floor(x) function may be useful, which returns the greatest integer less than or equal to the float x.

Test the program for the input π and $n = 20$.

Differential Equations

Exercise 8.5 The package DifferentialEquations also provides a BVProblem function, with which boundary value problems can be solved with the so-called *shooting method*, as explained in en.wikipedia.org/wiki/Shooting_method.

Consult the documentation and solve the Bratu equation using BVProblem with an appropriate shooting method, such as Shooting(Tsit5()).

Classes vs. Structs

Exercise 8.6 In the Python chapter we developed a class to represent polynomials, and then also a subclass for parabolas of degree 2. Recast the class Polynomial in Julia, using structs and the multiple dispatch feature.

As an advanced challenge, try to reformulate the concept of inheritance, using Julia's type hierarchy.

Spline Interpolation

In the following exercise we assume a basic knowledge of *spline interpolation*. The interested reader is referred to Sect. 10.5 in the Maple chapter for an introduction.

Exercise 8.7 Write a Julia program to interpolate the *Runge function*

$$f(x) = \frac{1}{1 + 25x^2}, \quad x \in [-1, 1],$$

using cubic splines. This requires the installation of an external package such as `Interpolations`, which contains a generator `CubicSplineInterpolation(x,y)` for cubic splines over equidistant mesh points.

Given a range x of e.g. 7 points, determine the associated y-values. Plot the resulting spline-function together with f.

Part III
Commercial Computing Environments

Chapter 9
MATLAB

MATLAB is a commercial programming environment, distributed by The MathWorks company, for numerical processing of mathematical problems, and graphical representation of solutions. The development began in the late 1970s as *Mat*rix *Lab*oratory, to make the linear algebra libraries that were written in the Fortran language available also to non-programmers. In the meantime MATLAB has developed into a universal program system for use in all areas of scientific computing.

MATLAB is available for the Linux, Windows and Mac operating systems, among others. Students can get cheaper licenses at `mathworks.com`. This site also offers extensive tutorials for free download.

With *GNU Octave* there is also a free software available which is largely compatible with MATLAB. Note however that the PDE toolbox discussed in Sect. 9.8 is not available in Octave.

A good introduction to MATLAB is provided in the free e-book [9]. For further reading the books [1] and [10] can be recommended.

9.1 Basics

MATLAB programs are usually run in its dedicated interpreter environment. The main component of the user interface is the "Command Window", in which you can enter a command at the prompt '>>'. It will then be executed immediately.

Example:

```
>>   2 * 4   % ans = 8
```

The variable `ans` corresponds to the anonymous variable '_' in Python. It stores the last output, which can then be accessed in reading mode. Comments are marked with the symbol '%'.

Note that, as in the previous chapters, we use a writing convention to save vertical printing space. We often write the result on the same line as the input, separated by the comment symbol.

© Springer Nature Switzerland AG 2020

E. Smith, *Introduction to the Tools of Scientific Computing*, Texts in Computational Science and Engineering 25, https://doi.org/10.1007/978-3-030-60808-8_9

As in Python, statements in Matlab do not have to be terminated by any special symbol. If a statement is terminated with a semicolon, the output is suppressed. A comma may be used to separate statements within a line.

We collect some useful little helpers. With `clc` you can empty the command window, with `clear` remove all variables in the workspace, with 'clear a' delete a single variable a. Clearing variables before a new computation can be very useful to avoid undesired side effects.

To stop execution of a Matlab command, for instance if you get stuck in a non-terminating loop, you can press `Ctrl+C` or `Ctrl+Break`. On Apple Macintosh platforms, you also can use `Command+.` (the Command key and the period key).

Typing `help` gives you basic information about a command, such as 'help sin' or 'help clear'. More detailed explanations can be obtained by entering `doc` instead of `help`.

Numbers and Arithmetic

Numbers can be of different types, e.g. `int32` for 32-bit integer numbers or `double` for 64-digit floating-point numbers. If no type is specified, Matlab will automatically assume `double`.

With regards to the basic arithmetic operators, it should only be noted that '^' denotes the power operator.

A list of the provided data types can be obtained with 'help datatypes'.

With 'format' the number representation can be changed. 'format short' is preset by default, which leads to an output of `sqrt(2)` as 1.4142, for example. After setting 'format long' it becomes 1.414213562373095.

Variables

The assignment of values to variables is carried out as in all previously considered languages with the assignment operator '='.

As in Python, a variable has no particular data type.

Boolean Operators, Comparisons

The type `logical` contains the truth values `true` and `false`. In the output, the truth values are represented as 1 and 0, whereby it is noted that they stand for logical values.

Boolean expressions can be composed with the operators '&' for *and*, '|' for *or*. Negation is indicated by '~'.

Comparisons are denoted by '<' for 'less than', '<=' for 'less than or equal', equality and inequality by '==' and '~='.

9.2 Vectors and Matrices

As mentioned, MATLAB was originally developed for matrix computations in linear algebra. This heritage is still visible throughout. For instance, if we make the assignment

```
>> x = 1;
```

and inspect the properties of the variable x with 'whos', we get

```
>> whos x    % returns property list
   Name     Size     Bytes    Class      Attributes
   x        1x1          8    double
```

The property 'Size 1x1' reveals a basic concept in MATLAB. All variables are conceived as matrices, a single number consequently as a 1×1 matrix. The property 'Class double' confirms our remark above that double is automatically assumed if no type is given.

Vectors

A vector is a sequence of numbers that is generated, for example, in the bracket notation

```
>> v = [1 2 3 4]   % v = 1 2 3 4
```

Entries are separated by a space or comma. They are accessed in the form v(i), with *indices starting at 1* as in Julia, as is usual in mathematics, and not at 0 as in Python or C/C++. The length of the vector is returned by length(v).

The Colon Operator and linspace

Using the *colon operator* ':', the input v = [1 2 3 4] can be shortened to v = 1:4.

The colon operator is one of the most important operators in MATLAB. With its help, for example, special vectors can be generated, used for indexing in loops or in plotting.

The usage is as follows: beginning with an initial number, a fixed number is added and stored in the vector until a predetermined end has been reached or exceeded. The general syntax is: '<start>:<increment>:<end>', or '<start>:<end>' assuming the default value 1 for the step size <increment>. Another example should make that clear:

```
>> x = 1: -0.3: 0   % x = 1.0000 0.7000 0.4000 0.1000
```

The colon operator is related to the Python function range(start,stop,step), where the latter however, generates numbers up to, but not including, the upper bound stop.

If instead of specifying the increment, you want to generate a *number n* of points of equal distance between values *a* and *b*, you use the function linspace(a,b,n), just as in NumPy.

Example:

```
>> linspace(-1, 1, 5)  % ans = -1.0000 -0.5000 0 0.5000 1.0000
```

The default for *n* is 100, so that linspace(a,b) is short for linspace(a,b,100).

Dynamic Vector Extension

For a given vector

```
>> v = [1 2]
```

we can simply generate an extension by assigning a value at some new index position:

```
>> v(5) = 1  % v = 1 2 0 0 1
```

The entries filled in at the new intermediate positions are automatically 0 initialized.

In fact, we can also *create* a vector in this way:

```
>> w(3) = 1  % w = 0 0 1
```

Comparison Operators and Logical Arrays

The comparison operators defined above can also be applied to vectors elementwise, whereby a *logical array* is returned:

```
1 >> x = [-1 2 3];
2 >> x > 0  % ans = 1×3 logical array  0  1  1
```

A logical array is an array with entries 1 and 0, which are, however, explicitly specified as Boolean values.

The logical array returned in line 2 can also be defined directly as follows:

```
>> b = [false true true]  % 1×3 logical array 0  1  1
```

Logical arrays can be composed with Boolean operators:

```
>> x = [true false true]; y = [true true false];
>> ~x     % 1×3 logical array  0  1  0
>> x & y  % 1×3 logical array  1  0  0
>>> x | y % 1×3 logical array  1  1  1
```

Remark In the introduction to this section, we noticed that a variable holding a number is automatically conceived as a 1×1 matrix. However, also a vector of length 1 is considered as a 1×1 matrix. In fact, a vector of length 1 is identified with its content in Matlab:

```
>> x = 1; y = [1];
>> x == y  % ans = logical 1
```

Matrices

Matrices can be generated explicitly with the bracket notation:

```
>> A = [1 2 3; 1 1 1; 3 3 1]
A =   1   2   3
      1   1   1
      3   3   1
```

The rows of the matrix are separated by a semicolon or a line break, the row entries by a space or comma.

Note that this confirms that the vectors considered above are single-row matrices. A column vector can be defined with entries separated by a semicolon, for example 'v = [1;2;3]'.

Matrix entries are accessed by A(i,j), with *indices starting at 1*.

Like vectors, matrices can also be defined and extended dynamically:

```
>> B(2,2) = 1
B =   0   0
      0   1
>> B(1,4) = 2
B =   0   0   0   2
      0   1   0   0
```

The Colon Operator for Matrices

As in NumPy, every subblock in a matrix can be accessed, not just individual elements. It is sufficient to replace the numbers with vectors when indexing.

The colon operator is often used to create a submatrix A(p:q,r:s) from a matrix A, which consists of the intersection of the rows p to q and the columns r to s.

Example:

```
>> A = [1 2 3 4; 5 6 7 8; 9 10 11 12; 13 14 15 16]; A(3:4,2:3)
ans =   10   11
        14   15
```

A special case is a single colon, which selects all rows or columns. For instance, A(:,j) denotes the *j*th column, and A(i,:) the *i*th row of *A*.

Standard Matrices

As in NumPy, many standard matrices can be generated directly with a MATLAB function. The *zero*, the *identity* and *one matrix* can be created using the functions zeros, eye and ones. All have the same syntax. For example, zeros(n,m) or equivalently zeros([n,m]) will produce an $n \times m$ zero matrix, while zeros(n) will produce a *square* $n \times n$ zero matrix.

Remark Note the slight difference between MATLAB and NumPy. In MATLAB a function with only one argument, say zeros(n), returns a *square matrix*, whereas in NumPy it returns a *vector*.

The function 'rand' is used to create matrices whose entries consist of pseudorandom numbers in the interval $[0, 1)$. The syntax is the same as above. Without argument, rand returns a single random number.

Of course, we can also use all the functions above to create special *vectors*, such as

```
>> ones(1,3)  % row vector
>> ones(3,1)  % column vector
```

Note again that the notation is slightly different from that in NumPy.

Special Matrices

MATLAB provides functions to create various special matrices. An example is the *Hilbert matrix*, whose elements a_{ij} have the value $1/(i+j-1)$. The matrix is generated by the command hilb(n) and its inverse by invhilb(n), e.g.:

```
>> invhilb(3)
ans =      9    -36     30
         -36    192   -180
          30   -180    180
```

By the way: This illustrates one of the remarkable properties of Hilbert matrices: The inverse consists only of whole numbers.

Over fifty other special and famous matrices can be generated with the command gallery. Which ones are available, can be inquired by 'help gallery'.

Matrix Operations

The usual matrix operations are available. The transpose of a matrix A is denoted by transpose(A) or A'. Matrix multiplication is computed by A*B, the *n*th power by A^n, and the inverse of a regular matrix A by inv(A).

We can also use operations between matrices and scalars. For example, 'A + n' adds the number n to each entry of the matrix A.

Division

One of the most powerful matrix operations in MATLAB is *matrix division*.

The slash symbol '/' denotes "*division from the right*". For matrices A, B (of compatible shape), A/B yields the matrix C, such that C*B returns A.

Example:
```
>> A = [1 2; 3 4]; B = [5 6; 7 8]; C = A / B
ans =   3.0000   -2.0000
        2.0000   -1.0000
>> C*B
ans =   1   2
        3   4
```

Essentially, A/B corresponds to the multiplication from the right with the inverse inv(B) of B.

Actually, the inverse of an $n \times n$ matrix B can be computed with eye(n)/B.

Dually, the *division from left* B\A yields the matrix C such that B*C returns A.

Componentwise Operations

The operations above can also be used componentwise. However, to avoid ambiguities we must precede the operator symbol with a dot '.':
```
>> [1 2; 3 4] .* [5 6; 7 8]
ans =    5   12
        21   32
```

Logical Matrices

As with vectors, a comparison like 'A < B' for two matching matrices results in a Boolean matrix with the truth values of the componentwise comparisons as entries.

In fact, a comparison between a scalar number and a matrix also returns a Boolean matrix:
```
>> A = [1 3; 2 0];
>>> 1 < A
ans = 2×2 logical array
   0   1
   1   0
```

Boolean matrices can also be composed with the logical operators '&', '|' and '~'.

A versatile method for the construction of matrices is to use Boolean matrices as filters. Assume a numeric matrix A and a matching Boolean matrix B. Then the operation A(B) returns a column vector consisting of all A entries for which the corresponding B entry has the value true.

We discuss an example where all of the above operations occur:
```
1 >> A = [1 2; 3 4];
2 >> B = [false true; true false]
3 >> A(B)  % ans = 3
4                  2
```

In line 2 B is defined as a Boolean matrix with the same shape as A. In line 3 the A entries corresponding to the value true in B are filtered out and represented as a column vector.

Line 5 tests the entries in A, whether their square is > 8. In lines 6–8 the answers are collected in a Boolean matrix:

```
5 >> A.^2 > 8
6 ans = 2×2 logical array
7    0  0
8    1  1
```

The term 'B | (A.^2 > 8)' in line 9 yields a matrix of truth values, obtained by componentwise disjunction of the two Boolean matrices B and (A.^2 > 8). The outcome is the logical array [0 1; 1 1]. The latter is in turn used to filter out all A entries except A(1,1), to each of which then the assignment '= 5' is applied. The result is shown in lines 10 and 11:

```
 9 >> A(B | A.^2 > 8) = 5
10 A =  1  5
11       5  5
```

9.3 Control Structures: Branching, Loops

Control structures are again very similar to the ones in Python and Julia.

Conditional Statements

The basic form of a conditional statement in MATLAB is

```
if   <condition>
     <statements>
else
     <statements>
end
```

The else-branch can again be omitted.

With elseif a conditional statement can be extended to

```
if <condition>, <statements>
elseif <condition>, <statements>
else <statements>, end
```

In conditional statements all components can be written in the same line, but then separated by a comma, or a semicolon if the output is to be suppressed.

Example:

```
>> if x > y, tmp = y; y = x; x = tmp; end
```

swaps the values of x and y, if $x > y$.

Loops

The for loop in MATLAB is a pure counting loop. The general form is

```
for <start>: <increment>: <end>, <statements>, end
```

If the increment is constant 1, the '<increment>:' part can be omitted.

Example 9.1 We define a tridiagonal 3×3 matrix with 4s on the main diagonal and -1s on the first super- and subdiagonals (actually the matrix B used in the construction of the Poisson matrix at the end of Sect. 4.1 in the SciPy chapter):

```
>> n = 3;
>> for i = 1:3, for j = 1:3
       if i == j, B(i,j) = 4;
       elseif abs(i-j) == 1, B(i,j) = -1;
       else B(i,j) = 0;   % redundant
   end, end, end
B =   4     -1      0
     -1      4     -1
      0     -1      4
```

Note again that B is generated dynamically. No previous shape declaration is required. As explained in Sect. 9.2, this also implies that the assignment in the else-line in the definition of B is actually redundant.

In addition to the for loop there is also a while loop of the form

```
>> while <condition>, <statements>, end
```

Example 9.2 (Machine Epsilon) As in Sect. 3.4 in the Python chapter, the following program computes the *machine epsilon*, i.e. the smallest positive floating point number ϵ for which $1 + \varepsilon > 1$:

```
>> epsi = 1;        % # spelling to avoid conflict
>> while 1 + epsi > 1, epsi = epsi/2; end
>> format long
>> epsi = 2*epsi    % epsi = 2.220446049250313e-16
```

The computed value matches the built-in MATLAB constant eps.

Remark As in Python and C, a forced loop termination is achieved by the command break in the body of the loop. The loop ends immediately, and control jumps to the line after 'end'. In contrast, the command continue aborts the *current* loop execution, and the loop continues with the next iteration.

Loops vs. Vectors

MATLAB is a language specifically designed for matrices and vectors. In particular, this implies that matrix and vector operations are implemented to perform efficiently.

Loops, on the other hand, require significantly more computing in MATLAB, hence direct operations should be preferred.

For illustration we create a value table y of length 1001 from sin(t), first using a loop construction

```
>> tic, i = 0;
for t = 0: 0.01: 10
  i = i + 1;
  y(i) = sin(t);
end, toc
```

and then, for comparison, by direct application of sin to the vector t:

```
>> tic, t = 0: 0.01: 10; y = sin(t); toc
```

The second version will run much faster, on our test machine by the factor 8.77.

On this occasion we also encounter the command pair tic and toc, which measures the time that has elapsed in the enclosed command block.

Note the different meanings of the expression 't = 0: 0.01: 10' in the two code snippets. In the first, the *number* t proceeds from 0 to 10 in steps of length 0.01. In the second, the variable t denotes a *vector* of the form $t = (0, 0.01, 0.02, \ldots, 9.99, 10.0)$, to which the sine function is then applied elementwise.

Elementwise function application will be explained more generally below.

9.4 Functions

The definition, computation and application of functions is a central concept in MATLAB. Somewhat exaggerated, you could say that programming in MATLAB is mainly about writing and evaluating functions.

In fact, the above mentioned operators are in reality nothing else than functions, only in special spelling.

Generally, a function in MATLAB does not change input arguments. Arguments are passed by value, not by reference.

Scalar or Elementary Functions

Many MATLAB functions are scalar functions and are executed elementwise when they are applied to matrices. These include the trigonometric, logarithmic and exponential functions. You can query what is available with the command 'help elfun'. You can get details on the individual functions with 'help sin' or more detailed information with 'doc sin'. The command 'help specfun' outputs a list of other special mathematical functions.

By clicking the *fx* symbol to the left of the current entry line you can access the total stock.

Scalar functions are often applied to vectors, as the following example shows:

```
>> x = linspace(0, 2*pi); y = cos(x.^2); plot(x,y)
```

First a vector x is generated with 100 equidistant values in the interval $[0, 2\pi]$. The y-vector then stores the cosine of the squares of the individual x values. Here again the dot in '.^' specifies that the power operator is to be applied elementwise. This is not necessary for the cosine function, since it is only defined for numerical values anyway.

The function plot should be self explaining. The output is shown in Fig. 9.1.

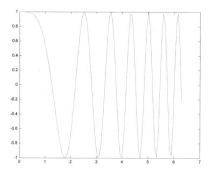

Fig. 9.1 MATLAB plot of $\cos(x^2)$ over the interval $[0, 2\pi]$

Vector Functions

A second class of MATLAB functions are the vector functions. They can be applied to both row and column vectors, using the same syntax. These functions include max, min and sum and prod, which compute the sum or product of the entries. length returns the length of a vector.

Matrix Functions

The strength of MATLAB are the matrix functions. Some important ones are: rank to query the rank of a matrix, det for the determinant, inv for computing the inverse. length returns the larger of the two dimensions for a matrix, numel the number of elements. size returns the dimensions. A possible usage of the latter is

```
>> [n,m] = size(A)
```

which stores the number of rows in n and that of columns in m.

The command diag(A) returns the diagonal of A (as a column vector). For a column vector v, conversely, diag(v) creates a diagonal matrix with the elements of v in the diagonal.

sum(A) creates a row vector, in which the elements of the single columns are summed up.

Here is a list of some other useful commands:

```
>> sum(diag(A))    % sums up the diagonal elements
>> sum(A')         % elements in the rows are summed
>> sum(A(:,3))     % sums up the elements of the 3rd column
>> sum(sum(A))     % sums up all elements of A
>> diag(diag(A))   % diagonal matrix
>> A(:,i) = []     % erases the i-th column from A
```

To reshape a matrix we use the reshape function:

```
>> A = [1 2 3 4; 5 6 7 8; 9 10 11 12; 13 14 15 16]
>> reshape(A, 2, 8)
ans =  1     9     2    10     3    11     4    12
       5    13     6    14     7    15     8    16
```

Note that reshape always assumes a column-major order. To obtain a row-major representation we can use the *transpose* operator as follows:

```
>> reshape(A', 8, 2)'
ans =  1     2     3     4     5     6     7     8
       9    10    11    12    13    14    15    16
```

9.5 M-Files

We have seen that Matlab provides a powerful interactive programming environment. Many procedures, which have to be elaborately formulated in other languages, are supplied as ready-made functions.

But also in Matlab you can save program components to files. However, this serves the main purpose to provide them as additional macros in command-line based programming.

Matlab typically uses so-called *M-files* for storage. The name comes from the simple fact that they are saved with the suffix '.m'. There are two types of M-files, *script files* and *function files*.

Script Files

Script files contain command sequences that are simply read into the system, then executed as if they had been typed directly. There are no input or output parameters. Script files are used when you have long command sequences, or when you want to avoid repeated typing of frequently used command blocks.

For the development of script files Matlab offers a comfortable text editor, which you can be access via the window bar or with the command 'edit <filename>' or simply 'edit'.

Example 9.3 We create a file scriptdemo.m with the content

```
disp('Generates 2 vectors of n random numbers')
n = input('n = ?: ');  % requests input n
x = rand(n,1); y = rand(n,1);
```

From the command line you can then inspect the contents of the file with

```
type scriptdemo
```

The script itself is used like this:

```
1 >> scriptdemo
2 Generates 2 vectors of n random numbers
3 n = ?:  7
4 >> [x y]'  % output result:
5 ans =  0.9501  0.2311  0.6068  0.4860  0.8913  0.7621  0.4565
6        0.0185  0.8214  0.4447  0.6154  0.7919  0.9218  0.7382
```

In line 3 the program asks for an input number, and receives 7 from the user. The result is then printed to the screen as a column of two row vectors.

Function Files

Function files store self-written functions, each function into its own file. Function files can be recognized by the fact that the first line of the M-file contains the word function. Functions are M-files with input and output parameters. The name of the M-file and the function should coincide. In case of doubt, MATLAB uses the file name.

To illustrate this, we define a function mv to compute the *mean* of an input vector and save it to a file mv.m:

```
%%% This function calculates the mean of a vector
function m  = mv(x)
n = length(x);
m = sum(x)/n;
```

Back in the command window, we can display the contents of the file by 'type mv'. The comment line is output with the command 'help mv'.

The function is applied as if it were a normal built-in function:

```
>> x = [1 2 3 4 5 6];
>> mv(x)  % ans = 3.5000
```

Subfunctions

Every function that is to be accessed by the user, must be in a file of its own. Subfunctions that are only called from *within* the main function can however be in the same file. Let's take a look at the following example function stat, stored in the file stat.m:

```
1 function [m,s] = stat(x)
2 n = length(x);
3 m = avg(x,n);  % subfunction defined below
4 s = sqrt(sum((x - m).^2)/n);
```

Note again that we precede the power operator in line 4 with a dot '.', since the operation is to be applied componentwise.

In line 3 a subfunction avg is called, which is defined in the same file:

```
5 function m = avg(x,n)
6 m = sum(x)/n;
```

The function stat can be called like this:

```
>> [mv, stddev] = stat(x)
```

However, we cannot use avg directly because it is in the file stat.m.

Functions as Arguments

A function can also be passed as argument to another function.

As an example, consider the following plotfunc function that plots an input function f over a vector x of evaluation points:

```
function plotfunc(f,x)
y = feval(f,x);
plot(x,y);
```

The MATLAB command feval evaluates the function f at the components in x.

The function plotfunc is used as follows:

```
>> x = -3: 0.01: 3;
>> plotfunc(@sin, x);  % note the prefix @
```

This will draw the sine function in the interval $[-3, 3]$. Note that the input function must be preceded by the special *function handle* symbol '@'.

Anonymous Functions

The only exception to the rule that functions must be stored in files of their own, is that so-called *anonymous functions* (corresponding to lambda functions in Python) may be defined in the command window.

Example:

```
1 >>  f = @(x) x.^2 + 2*x;  % @ creates function handle
2 >>  f(.5)                 % ans = 1.2500
```

The expression in line 1 defines f to be the function $x \mapsto x^2 + 2x$. The @ operator creates a function handle, and the parentheses immediately after the operator include the function input arguments.

Line 2 shows that the result behaves like a normal function.

Remark Note that we have written the power operator '.^' in the form for componentwise application. If we know that the function will only be applied to single scalar values x, the dot is redundant (but still perfectly legal), and we can define it like this:

```
f_sc = @(x) x^2 + 2*x  % not applicable to vectors
```

But the dot *does* matter if we want to plot the function, say

```
>>  x = linspace(0,1);
>>  plot(x, f(x))
```

We can also define anonymous functions with multiple variables, such as

```
>> g = @(x,y) x.^2 - y.^2
```

Save and Load Files

Also in MATLAB there are many ways to save and read files. With

```
>> A = [1 2; 3 4];
>> save('a.txt', 'A', '-ascii')
```

the matrix A is coded as a character string and written row by row to the file a.txt in the MATLAB default directory. Without the option '-ascii', the data is stored in a binary format.

To read the data into a matrix B, we use the command

```
>> B = load('a.txt');
```

For details and further possibilities see 'help save' and 'help load'.

9.6 Linear Algebra

Based on the built-in standard operators and functions, we can already carry out extensive matrix computations. We consider some examples of linear equation systems and matrix decompositions.

Linear Equation Systems

MATLAB offers several possibilities to solve linear equation systems. The simplest and most versatile one uses the backslash operator "division from the left" introduced above.

Example 9.4 Let

```
>> A = [1 2 3; 1 1 1; 3 3 1]; b = [2 2 0]';
```

The system of linear equations $Ax = b$ is then solved by

```
>>  x = A \ b
x =   5.0000
     -6.0000
      3.0000
```

Hidden behind the backslash operator is the most important method for solving systems of linear equations, Gaussian elimination based on LU decomposition. The decomposition is however used only in the course of the calculation, and is afterwards no longer available.

Below we show how to explicitly generate LU decompositions in MATLAB.

Matrix Decompositions

MATLAB offers a large collection of matrix factorizations, among them various forms of LU-, Cholesky-, and QR-decompositions.

LU Decomposition

Recall that the algorithm factorizes a nonsingular square matrix A into a unit lower triangular matrix L with only 1-entries in the diagonal, an upper triangular matrix U and a permutation matrix P, such that $A = PLU$.

In Sect. 4.4 in the SciPy chapter we used the function lu to generate such a decomposition. In MATLAB we have a function with the same name, however a slightly different behavior. It does not return the permutation matrix P such that $A = PLU$, but rather the inverse such that $PA = LU$:

```
>> A = [1 2 3; 1 1 1; 3 3 1];   [L,U,P] = lu(A)
L =   1.0000        0        0
      0.3333   1.0000        0
      0.3333        0   1.0000
U =   3.0000   3.0000   1.0000
           0   1.0000   2.6667
           0        0   0.6667
P =   0   0   1
      1   0   0
      0   1   0
```

Cholesky Decomposition

Recall that the *Cholesky decomposition* factorizes a *symmetric positive definite* matrix A into a product $U^T U$ where U is an upper triangular matrix.

In MATLAB the Cholesky decomposition is computed by

```
>> A = [1 2 1; 2 5 2; 1 2 10];  U = chol(A)
```

with the same result as in the corresponding SciPy example in Sect. 4.4.

QR Decomposition

Recall that the QR decomposition of a matrix A is a factorization $A = QR$ into an orthonormal matrix Q and an upper triangular R. In MATLAB, QR decomposition is computed by the function qr, with the same behavior as in SciPy:

```
>> A = [12 -51 4; 6 167 -68; -4 24 -41]; [Q,R] = qr(A)
Q =    -0.8571     0.3943     0.3314
       -0.4286    -0.9029    -0.0343
        0.2857    -0.1714     0.9429
R = -14.0000   -21.0000    14.0000
           0  -175.0000    70.0000
           0          0   -35.0000
```

Note that in MATLAB the "economic" form of the qr function, which we used in the least square method in SciPy, is still readily available:

```
>> A = [3 -6; 4 -8; 0 1]; b = [-1 7 2]';
>> [Q,R] = qr(A,0)   % second arg 0 specifies economic form
Q =    -0.6000         0
       -0.8000         0
             0   -1.0000
R =   -5   10
       0   -1
>> x = R\Q'*b   % x = [5, 2]'
```

Remark In fact, the least-squares solver is already available directly in MATLAB. The backslash operator can also be used for this purpose:

```
>> A = [3 -6; 4 -8; 0 1]; b = [-1 7 2]';
>> x = A\b   % x = [5.000, 2.000]'
```

9.7 Ordinary Differential Equations

We have seen that for historical reasons matrix computations are a strength of MATLAB. As mentioned earlier, powerful methods are now available for many other fields of applications. This is not least the case for processing differential equations, especially ordinary first-order equations ODEs.

In Sect. 4.7 in the SciPy chapter we discussed various examples of SciPy programs for the solution of ODEs. We indicate how corresponding programs can be written in MATLAB. For background we refer to the discussion in the SciPy chapter.

As a first example we recall the *initial value problem*

$$
(1) \qquad \begin{aligned} u'(x) &= 2x, \quad x \in [0,1], \\ u(0) &= 0, \end{aligned}
$$

with the exact solution $u_e(x) = x^2$.

In MATLAB the equation can for instance be solved by direct integration according to the Euler method like this:

```
u(1) = 0;  % indices begin at 1
n = 100;
x = linspace(0, 1, n);
for i = 1:(n-1) u(i+1) = u(i) + 2*x(i)/n; end
plot(x,u)
```

The MATLAB Function ode45

However, for first-order ODEs like (1), MATLAB offers highly efficient built-in solution modules, in particular the solver function ode45. The first part of the name obviously refers to ordinary differential equations. The '45' is to indicate that it implements a special type 4/5 Runge-Kutta method, the *Dormand-Prince* method.

Remark Actually, the SciPy solver solve_ivp in Sect. 4.7 in Chapt. 4 uses precisely this method as default.

Our ODE (1) above can then be solved like this:

```
>> dudx = @(x,u) 2*x;  % anonymous function def, see above
>> [x,u] = ode45(dudx, [0, 1], 0);
>> plot(x,u)
```

Here dudx specifies the derivative u', the second argument specifies the solution interval, the third the initial value $u_0 = 0$.

Initial Value Problems

We use ode45 to solve the initial value problem

$$u'(x) = x - u(x), \quad u(0) = 1, \quad x \in [0, 5]$$

from Sect. 4.7 in the SciPy chapter:

```
>> dudx = @(x,u) x - u;
>> [x,u] = ode45(dudx, [0, 5], 1);
```

Note that ode45 iteratively modifies the vector x of evaluation points until a predetermined precision is reached.

This can be seen in the solution plot in Fig. 9.2 when we include the computed nodes:

```
>> plot(x,u,'-o')
```

Second Order ODEs

We illustrate how to solve second order ODEs with ode45. The example below corresponds to the pendulum example in the SciPy chapter, but since the techniques are of general importance, we discuss it in detail.

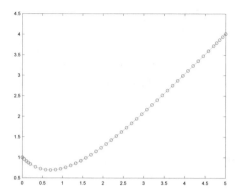

Fig. 9.2 ode45 solution of $u'(x) = x - u(x)$, $u(0) = 1$ over the interval $[0, 5]$. The circles indicate the approximation grid chosen by the solver function

We consider a simple form of an oscillation equation with the solution $u = \sin$:

(2)
$$u'' + u = 0$$
$$u(0) = 0, \ u'(0) = 1.$$

ode45 can only deal with first order ODEs. We use the standard technique and transform (2) into a system of *two* first order equations.

We define $v := u'$ and get the system

$$v = u',$$
$$v' = -u,$$
$$u(0) = 0, \ v(0) = 1.$$

This system can now be processed by ode45, if we pass it over in a suitable form.

We collect u and v in a column vector

$$y = \begin{pmatrix} u \\ v \end{pmatrix} \quad \text{such that} \quad \frac{d}{dx} y = \begin{pmatrix} y_1' \\ y_2' \end{pmatrix} = \begin{pmatrix} u' \\ v' \end{pmatrix} = \begin{pmatrix} v \\ -u' \end{pmatrix} = \begin{pmatrix} y_2 \\ -y_1 \end{pmatrix}.$$

This rightmost pair of derivatives is passed to MATLAB in the form

```
1 dydx = @(x,y) [y(2) -y(1)]';
```

Note that the specification as *column* vector is required.

We store the initial values $u(0)$, $v(0)$ in a vector

```
2 y0 = [0 1];
```

and fix an interval for the evaluation, say

```
3 xint = [0 10];
```

With these input parameters we call ode45:

```
4 [x y] = ode45(dydx, xint, y0);
5 u = y(:,1);
```

The solution vector y contains both solution functions u and v. In line 5 we choose the one we are interested in. The solution can then be plotted with plot(x,u).

Boundary Value Problems

As stated in the SciPy chapter, boundary value problems BVPs are generally more difficult to deal with than initial value problems. This also applies to Matlab.

Consider the following BVP, discussed in Chapt. 4:

$$u'' = -2,$$
$$u(0) = 0, \ u(5) = 3.$$

To process it with Matlab, we first translate it to a system of two first order equations:

```
1 dydx = @(x,y) [y(2) -2]';
```

We then define a function to compute the boundary conditions:

```
2 bc = @(lb,rb) [lb(1) rb(1)-3]';   % index '1' refers to u = y(1)
```

The boundary values are stored in residual form: "What residuum needs to be subtracted to get a homogeneous equation?".

Similar to the Python function solve_bvp, the Matlab solver needs initial values for the solution process. These are handled by the built-in function bvpinit.

The first argument in bvpinit takes an initial mesh for the solution interval. In our case it is sufficient to provide the endpoints of the interval, i.e. the vector [0 5].

The second argument stores a guess for the solution. In our case, the vector [0 0] instructs the system to start with the constant value 0 for all mesh points for both functions y(1) and y(2).

These initial assumptions are assigned to a variable solinit:

```
3 solinit = bvpinit([0 5], [0 0]);
```

Now the special Matlab solver bvp4c can be applied:

```
4 sol = bvp4c(dydx, bc, solinit);
```

The variable sol contains the solution and additionally a detailed protocol of the solution process.

We want to plot the solution function u. To this end we fix

```
5 x = linspace(0,5);   % recall: 100  equally spaced points
```

and then call the Matlab function deval to extract the function values as follows:

```
6 y = deval(sol, x, 1);
7 u = y(1,:);
```

In line 6 the value matrix for both y(1) and y(2) is returned, from which we extract the row representing the solution for the original function u in line 7.

The result can then again be plotted with plot(x,u).

The Bratu Equation

In Example 4.14 in the SciPy chapter we solved the Bratu equation

$$u''(x) = -e^{u(x)}, \quad x \in [0, 1],$$
$$u(0) = u(1) = 0.$$

Here is the MATLAB version of the solution program:

```
1 dydx = @(x,y) [y(2) -exp(y(1))]';
2 bc = @(lb,rb) [lb(1) rb(1)]';
3 meshinit = linspace(0, 1, 3);
4 solinit = bvpinit(meshinit, [0 0]);  % initial guess u = u' = 0
5 sol = bvp4c(dydx, bc, solinit);
6 x = linspace(0,1);
7 y = deval(sol, x, 1);
8 u = y(1,:);
9 plot(x,u);
```

Recall that the Bratu equation has *two* solutions. The program above returns the solution based on an initial guess $u(x) = u'(0) = 0$ for all mesh points. To generate the other, it is sufficient to replace the second argument of bvpinit in line 4 with [3 0], specifying that we now assume the initial estimate $u(x) = 3$ for all mesh points.

9.8 Partial Differential Equations

Similar to SciPy, MATLAB does not have any ready-made solver functions for general partial equations. (As we will see below, MATLAB does however provide a versatile graphical toolbox to help in the development of solutions.)

Here we discuss a MATLAB variant of the SciPy program developed for the solution of *2D Poisson equations* in Sect. 4.8 in Chapt. 4.

Recall that the basic case consisted in determining a function $u \colon \Omega \to \mathbb{R}$, $\Omega := [0, 1] \times [0, 1]$, such that

$$-\Delta u \equiv 1 \quad \text{in } \Omega,$$
$$u \equiv 0 \quad \text{on the boundary } \partial\Omega.$$

For background information we refer to the introduction in Sect. 4.8, and turn directly to the solution program, written as a MATLAB script.

We begin with the grid size:

```
1 n = 100;  % n x n inner grid points
2 h = 1/(n+1);
```

The value matrix, in which the solution is to be stored, is initialized with 0s:

```
3 u = zeros(n+2);  % recall: square matrix, unlike Python
```

The 0s at the boundary will retain that value, the inner components in u will assume the computed solution values.

Recall that the latter are determined by solving

$$\frac{1}{h^2} Au = b$$

with the $n^2 \times n^2$ Poisson matrix A and a 1-vector $b := (1, 1, \ldots, 1)^T$ of length n^2.

Here we are not constructing the Poisson matrix, like we did before; it is already available in the Matlab collection of special matrices:

```
4 A = gallery('poisson', n);  % built-in Poisson matrix A
5 b = ones(n*n, 1);           % needed as column vector
```

Note that A is provided as a sparse matrix. This can be seen as usual with 'whos A'.

The system is then solved, the solution stored in a column vector u_inner of length n^2:

```
6 u_inner = (A/h^2) \ b;
```

The column vector is then reshaped into an $n \times n$ matrix and inserted as a block of inner points into the solution matrix:

```
7 u(2:n+1, 2:n+1) = reshape(u_inner, n, n);  % column-major order
```

The solution can then be plotted with

```
8 lin = linspace(0, 1, n+2);
9 [x y] = meshgrid(lin);  % grid
10 mesh(x, y, u, 'LineWidth', 1);
```

Further elaboration and extension of the program is left to the exercises.

The Matlab PDE Toolbox

Matlab offers a powerful toolbox for the approximate solution of partial differential equations. As mentioned, a corresponding toolbox is unfortunately not available in Octave.

We illustrate the use for the following Poisson equation, discussed in Sect. 4.8 in the SciPy chapter:

$$-\Delta u = f = 1.25 \cdot e^{x + \frac{y}{2}} \quad \text{in } [0, 1] \times [0, 1],$$

$$u = g = e^{x + \frac{y}{2}} \quad \text{on the boundary:}$$

- By entering

  ```
  >> pdeModeler
  ```

in the command window we start the graphical user interface.

- In the menu item "Options -> Axes Limits" we change both axis limits to $[0, 1]$.

- By clicking on the rectangle symbol at the top left, we declare that our PDE should be evaluated over a rectangular area.

- By clicking on the point $(0, 0)$ with the left mouse button, and then, holding the button pressed, dragging the pointer to the point $(1, 1)$, we determine the size of the rectangle.

- By double-clicking on the rectangle, input fields appear, in which we can adjust the corner points.

- Next, we specify the boundary values. By clicking on the symbol $\partial\Omega$, the border lines are displayed in red. By clicking on the lines, the boundary conditions can now be set. For each boundary line we enter `exp(x+y/2)` for r, and leave the value $h = 1$ unchanged.

- By clicking on the PDE symbol, the equation itself can now be specified. In our case we keep the preselection "Elliptic" for the type of PDE and $c = -1, a = 0$. We set f to `-1.25*exp(x+y/2)`.

- By clicking on the 'Δ' symbol, we can display the grid, on which the solution function is approximated. Note that `pdeModeler` chooses a *triangular* grid pattern. This is because the underlying approximation method is based on the so-called *finite element method*. We return to the finite element method in later chapters.

- We click on the '$=$' symbol to compute the solution. In the menu item "Plot -> Parameters", the graphic representation of the solution can be controlled, for example by selecting the two options "Height (3-D plot)" and "Plot in x-y grid".

The graph then corresponds to the SciPy solution at the end of Sect. 4.8 in Chapt. 4.

9.9 Exercises

Matrix Computations

Exercise 9.1 Compute the regression line $g = a_1 x + a_2$ through the point cloud in $\mathbb{R}^2$, generated by

```
>> x = 0:20;
>> y = x + 2*rand(size(x));
```

Plot the line together with the cloud. To plot the scattered points you can use the `plot` command with a third argument, say `plot( , , '*')`. To ensure that both plots appear in the same figure, use 'hold on' after the first.

ODE, Initial Value Problem

Exercise 9.2 Use ode45 to solve the third order nonlinear initial value problem

$$u \cdot u''' = -1,$$
$$u(0) = 1, \ u'(0) = u''(0) = 0,$$

and plot the solution.

ODE, Boundary Value Problem

Exercise 9.3 Use bvp4c to solve the boundary value problem

$$u'' - 2u' + u = 0,$$
$$u(0) = u(1) = 0,$$

and plot the solution.

Partial Differential Equations PDEs

Exercise 9.4 In analogy to the SciPy function poisson_solver, extend the solution program in Sect. 9.8 to a MATLAB function to solve general Poisson equations

$$-\Delta u = f \quad \text{in } [0, 1]^2,$$
$$u = g \quad \text{on the boundary.}$$

Test the function for

(1) $$f = 1.25 \cdot e^{x + \frac{y}{2}}, \quad g = e^{x + \frac{y}{2}},$$

(2) $f = 20 \cos(3\pi x) \sin(2\pi y),$ $\begin{aligned} u(0, y) &= y^2, & u(1, y) &= 1, \\ u(x, 0) &= x^3, & u(x, 1) &= 1 \end{aligned}$ on the boundary.

Exercise 9.5 Solve the following Poisson equation over a *non-square* rectangle:

$$\frac{\partial^2 u}{\partial x^2} + \frac{\partial^2 u}{\partial y^2} = x^2 + y^2, \quad (x, y) \in [0, 2] \times [0, 1],$$
$$u(x, y) = x(2 - x) + y(1 - y) \quad \text{on the boundary,}$$

either directly, with a program as in the text, or by application of the pdeModeler in the MATLAB PDE Toolbox.

Advanced Topic: The Mandelbrot Set in Complex Numbers

In the exercises in the C++ chapter we looked at the set $\mathbb{H}$ of quaternions. Here we return to a more conventional number set, the complex numbers $\mathbb{C}$. In Sect. 8.9 in the Julia chapter, we discussed Julia sets, and mentioned that they are related to the probably better known Mandelbrot set. We follow up that topic here.

More precisely we study the behavior of the sequences $s(c)$ given by

$$z_{c,n+1} = z_{c,n}^2 + c, \quad z_{c,0} = 0, \quad \text{for } c \in \mathbb{C}.$$

Some examples: For $c = -2$ the sequence is stationary: $0, -2, 2, 2, \ldots$, for $c = 2$ it diverges: $0, 2, 6, 38, \ldots$, for $c = -i$ it alternates: $0, -1 - i, i, -1 - i, i, \ldots$.

A simple observation shows that once the absolute value $|z_{c,n}|$ of a sequence member is > 2, the sequence $s(c)$ will diverge.

The *Mandelbrot set* is the set

$$M = \{c \in \mathbb{C} \mid s(c) \text{ is bounded }\}.$$

Representing the complex numbers as $\mathbb{R}^2$, the Mandelbrot set can be visualized as in Fig. 9.3.

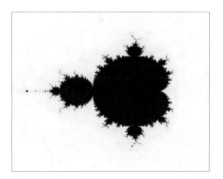

Fig. 9.3 The Mandelbrot set

Exercise 9.6 Plot an approximation to the Mandelbrot set.

We give a few hints to support the development.

A coarse calculation shows that $M \subseteq R := [-2, 1] \times [-1.5, 1.5]$. We start with the assumption that $M = R$. The idea is now to check for $c \in R$, whether $s(c)$ diverges, and thus establishes *non*-membership in M.

Clearly, in practical computations we must limit the check to a *finite* number of iterations. We thus only get an approximate solution.

Concerning the MATLAB implementation:

A complex number is written in the form a + b*i. The domain R should be represented by a discrete finite grid. We then need two matrices, one, say Z, for the storage of the current sequence values $z_{c,n}$, and one, say S, for the number of steps n needed to reach $|z_{c,n}| > 2$ (if possible). The matrix S can then be plotted with the function imagesc.

To allow interactive tuning of the solution, it is advisable to formulate the program as a MATLAB function with two parameters: the number of grid points, and the maximal number of iterations.

The matrix computations can of course be embedded in several for loops. However, with the sophisticated matrix operations provided by MATLAB, in particular the logical matrices discussed at the end of Sect. 9.2, the whole program will actually only require *one* for loop.

Chapter 10
Maple

Based on the same ideas as Symbolic Python, Maple (for *ma*thematical mani*p*ulation *l*anguage) is a commercial computer algebra system for interactive programming in algebra, analysis, discrete mathematics, graphics, numerical computations and many other areas of mathematics. As the name suggests, it was developed in Canada, from 1980 onwards at the University of Waterloo.

Initially focused on symbolic computing, Maple was over time developed into a powerful system also for numerical algorithms.

Maple works well with MATLAB and complements the procedures there with the ability to represent real numbers with any required (but still finite) accuracy. The book [4] provides a comprehensive discussion of scientific computing with the combination of MATLAB and Maple.

10.1 Basics

The main component of the Maple user interface is the "Worksheet" for interactive programming. In the command line, after the prompt '>' you can enter a Maple command, which will then be executed immediately. A command is usually completed by a *semicolon*. Entering a *colon* suppresses the output.

The central feature of all computer algebra systems is that they by default operate with *exact* values. Like Python, Maple can represent natural numbers of any size: 2^100 yields the exact result 1267650600228229401496703205376. Note that the power operator is denoted by '^'. For rational numbers, entering 115/39 + 727/119 returns the exact result 42038/4641.

The result of sin(Pi/4) will only be output as $1/2\ \sqrt{2}$, because an exact representation of the root is not possible. Here Pi obviously denotes the number π.

A numerical representation of symbolic values is obtained with evalf:

```
sin(Pi/4):
evalf(%);  # Out: 0.7071067810
```

© Springer Nature Switzerland AG 2020
E. Smith, *Introduction to the Tools of Scientific Computing*, Texts in Computational
Science and Engineering 25, https://doi.org/10.1007/978-3-030-60808-8_10

Note that as in Python, comments are indicated by the symbol '#'. As before we often write the result of an operation as a comment on the input-line.

The Maple symbol '%' denotes the so-called "ditto" operator, i.e. the content of the previous line. The name of the function evalf stands for *evaluate using floating-point arithmetic*. Maple is exact to a number of digits determined by the environment variable Digits, with default value 10.

Calling kernelopts(maxdigits) one obtains the maximal value for Digits, on the system used here 38654705646. The desired number of digits can also be passed to the function evalf as an additional argument, such as evalf(%,5) or equivalently evalf[5](%).

The root function sqrt may also serve to explain some further Maple features: sqrt(2) returns the expression $\sqrt{2}$, whereas the decimal point in sqrt(2.0) requests the output in decimal approximation, such as 1.414213562.

In many cases you can contribute to the calculation with additional help:

```
1 sqrt(x^2);
```

$$\sqrt{x^2}$$

```
2 assume(x, nonnegative):
3 sqrt(x^2);
```

$$x\sim$$

In line 1 the system gets stuck, because it does not know if x is a positive or negative value. The annotation in line 2 makes this clear. The tilde in the output marks that we make an additional assumption about x.

By default Maple performs calculations in the complex numbers $\mathbb{C}$. For example, sqrt(-1) returns the result I, the Maple symbol for the imaginary unit i. The term exp(I*Pi) + 1 evaluates to the correct result 0, thus confirming *Euler's identity*.

However, restriction to real-valued computation is possible.
    Example:

```
1 sqrt(-4);
```

$$2I$$

```
2 with(RealDomain):
3 sqrt(-4);
```

$$undefined$$

In line 2 the program package RealDomain is loaded, which among others has the effect that now sqrt is interpreted as the real-valued root function. Exactly which of the Maple components are modified in the package, is listed when the command with(RealDomain) is terminated with a semicolon instead of a colon. We will return to additional packages in more detail later.

With the '?' operator, you can open the help menu and consult the documentation, e.g. '?sqrt' for the root function.

Variables

Naturally, results can also be assigned to variables. Note, however, that the assignment operator is ': =' and not '=' as in the languages we have seen so far.

The assignment 'a := sin(Pi/4)' stores the symbolic expression $1/2 \sqrt{2}$ in the variable a. As before, we can then evaluate the expression numerically by evalf(a) or make an assignment a:=evalf(a), which then overwrites the original symbolic expression.

Values of variables are preserved as long as the worksheet is open. To avoid possible side effects, it is a good idea to reset the variables before new computations, e.g. by unassign('a','b','c') or by 'restart', which clears all entries in Maple's memory.

10.2 Functions

Maple has an extensive collection of mathematical functions, from the abs function for calculating the absolute value of real or complex numbers to the Rieman-Hurwitz ζ function zeta. The entire list can be obtained by '?initial'.

User Defined Functions

There are basically three ways to define functions in Maple, the arrow operator, the unapply command, and the proc definition.

The Arrow Operator

The arrow operator '->' corresponds to the mathematical maps-to operator '$\mapsto$'.

With it functions can be formulated in a form that corresponds to the lambda expressions discussed before, such as

```
1 f := x -> a*x^2;
2 g := (x,h) -> (f(x+h) - f(x)) / h;
3 delta := (x,y) -> if x = y then 1 else 0 fi;
```

The functions are then applied like the built-in functions: f(2) yields $4a$, g(2, 0.01) results in 4.010000000 a, and delta(I,sqrt(-1)) has the value 1.

Note that the expression 'if ... fi' in line 3 is the Maple syntax for the ternary operator '?:' encountered in C.

The unapply Command

With the help of the unapply command a term expression can be converted to a function. For example, consider the following expression:

```
term := x + x^2:
```

We can evaluate this term, for example with

```
eval(term, x = 1)   # Out: 2
```

However, the term is not yet a function in the sense of x -> x+x^2. This is what the unapply command is for:

```
f := unapply(term, x)
```

$$f := x \mapsto x + x^2$$

The name comes from the fact that a function is *applied* to an argument, and the result is then a *term*. The name unapply now suggests the converse. Recall that in SymPy the corresponding function was called lambdify.

The proc Definition

The proc definition (for procedure) corresponds to the *function* definitions in Python, C and Julia. It goes far beyond the definition method based on the arrow operator, since all language constructions can be used in a proc definition, including control structures such as if-statements and loops.

As a first (trivial) example, the function $f(x) := x^2$ can be defined as follows:

```
f := proc(x) return x^2; end;
```

The return statement gives the result value of the procedure:

```
f(2);   # Out: 4
```

If the procedure does not contain an explicit return statement, the value of the last term is automatically returned. Thus the above example could be formulated more concisely by

```
f := proc(x) x^2; end;
```

Here are some less trivial examples:

Example 10.1 (Factorial)

```
fact := proc(n)
    local i, prod;
    if n < 0 then return "n must be positive"; end;
    prod := 1;
    for i from 1 to n do prod := i*prod; end;
    return(prod);
end;
```

The keyword `local` specifies that the variables in the function definition are to be considered as different from possibly same-named variables in the worksheet environment.

Example 10.2 (Collatz Problem)

```
fcollatz := proc(n)
    local m, count;
    count := 0; m := n;
    while m > 1 do
        if modp(m,2) = 0  then m := m/2;
        else  m := 3*m + 1; end;
        count := count + 1; end;
    return count;  # steps required to reach 1
end;
```

Example 10.3 (Recursive Factorial)

```
fact_rec := proc(n)
    if n = 1 then return 1; end;
    return n*fact_rec(n-1);
end;
```

To get a feel for the possibilities of the `proc` definition, you can put Maple into a talkative mode with the command `interface(verboseproc=2)` and then inspect the definitions of built-in functions with e.g. `eval(nextprime)`.

Remark Here we refrain from detailed explicit syntax considerations of the control structures. They can be queried from Maple itself. For the sake of completeness, it should be mentioned that control structures can of course also be used outside of `proc` definitions.

Plotting

With the command `plot` Maple offers extensive options for the graphical representation of functions. The input of a function `f` must in general be provided as a function *term*, say, `f(x)`. In particular, if `f` is defined directly by a term, such as

```
f := x -> x^2:
```

then either of the forms `f(x)` or `x^2` can be used.

Fig. 10.1 shows a representation of `f` together with another function, which is given directly in the term form `x^3`:

```
plot({f(x), x^3}, x = 0..1, color = [red, blue]);
```

3-dimensional plots can be generated like this:

```
plot3d(x^2 + y^2, x = -1..1, y = -1..1);
```

With the command `with(plots)` many other plots options can be loaded, e.g. `animate` and `animate3d` for the representation of temporal processes.

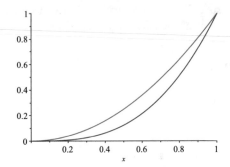

Fig. 10.1 Maple plot of function $f : x \mapsto x^2$ (red), and term evaluation x^3 (blue)

Equation Systems

Maple offers powerful methods for solving systems of equations, both *symbolic* with solve, and *numeric* with fsolve.

Symbolic:

```
solve(sin(x) = cos(x))
```
$$\frac{1}{4}\pi$$

```
quad := x^2 + 2*x + c = 0:
solve(quad, x);
```
$$-1 + \sqrt{1 - c}, \ -1 - \sqrt{1 - c}$$

Numeric:

```
eq1 := (sin(x + y))^2 + exp(x)*y + cot(x - y) + cosh(z + x) = 0:
eq2 := x^5 - 8*y = 2:
eq3 := x + 3*y - 77*z = 55:
fsolve({eq1, eq2, eq3});
```
$$\{x = -1.543352313, \ y = -1.344549481, \ z = -.7867142955\}$$

10.3 Linear Algebra

Maple knows the standard methods of linear algebra. Both symbolic and numerical calculations are supported. The LinearAlgebra package must be loaded in order to take full advantage of these capabilities. But first let's check how far we get with the standard on-board resources.

A vector is defined as follows:

```
v := Vector([1, 0, -3]):  # or equivalently:
w := <1, 0, -3>:
```

In both cases the vector is conceived as a column vector $(1, 0, -3)^T$. Components are accessed in the form v[i], with indices starting at 1, so that v[3] is -3.

We get a row vector like this:

```
v := Vector[row]([1, 0, -3]):
```

Remark Brackets can also be used as a subscript operator in general, e.g. to define a sequence a_1, a_2, a_3 by a[1], a[2], a[3]. However, this does not create a as a vector. From whattype(a) we get the answer *symbol*. For a genuine row vector v as above, whattype(v) gives the answer *Vector$_{row}$*.

Matrices can be entered as a list of row vectors:

```
A = Matrix([[1, 2], [3, 4]]);
```

or as a sequence of column vectors:

```
A := <<1, 3> | <2, 4>>;
```

The output is in both cases

$$A = \begin{bmatrix} 1 & 2 \\ 3 & 4 \end{bmatrix}$$

The entries are accessed by A[i,j], with indices starting at 1.

Matrix multiplication is denoted by the '.' operator, such as A.v or A.B. The shapes must of course match. In particular, the dot product v.w between vectors v and w is defined if the vectors have the same length.

Function Values as Matrix Entries

Maple provides a special mechanism for creating vectors and matrices, whose entries can be described by a function. Consider a vector v with entries i^2, $i = 1, \dots, 10$. It can be generated as follows:

```
f := i -> i^2:
v := Vector(10, f);
```

The syntax requires the length of the vector and a function that computes the *i*th component.

Also a short-hand notation is allowed:

```
v := Vector(10, i -> i^2);
```

If no function is specified, the value 0 is assumed by default. The assignment

```
v := Vector(3);
```

returns the zero vector $(0, 0, 0)^T$.

The same technique can also be used for matrices. Here is the Hilbert matrix:

```
H := Matrix(3, 3, (i,j) -> 1/(i+j-1), shape=symmetric);
```

$$\begin{bmatrix} 1 & 1/2 & 1/3 \\ 1/2 & 1/3 & 1/4 \\ 1/3 & 1/4 & 1/5 \end{bmatrix}$$

With 'shape=symmetric' we tell the system that H is a symmetric matrix. Such additional information is often useful, if – as here – it indicates that space-saving storage is possible. This additional information is also helpful when calculating the eigenvalues below.

Vector Space Bases

We recast some of the examples in Chapt. 5 from SymPy to Maple.

Example 10.4 In Example 5.2 we saw that the vectors

$$v_1 := \frac{1}{\sqrt{3}}(1,1,1), \quad v_2 := \frac{1}{\sqrt{2}}(1,0,-1), \quad v_3 := \frac{1}{\sqrt{6}}(1,-2,1)$$

form an orthonormal basis of $\mathbb{R}^3$. In Maple, the verification program looks like this:

```
1 v1 := <1, 1, 1> / sqrt(3):
2 v2 := <1, 0, -1> / sqrt(2):
3 v3 := <1, -2, 1> / sqrt(6):
4 v1.v1:
5 v1.v2:
6 x := <a,b,c>:
7 y:=(v1.x)*v1 + (v2.x)*v2 + (v3.x)*v3;
8 simplify(y)
```

The vector y returned in line 7 has the same unintelligible form as the one obtained from SymPy. Again, it is the `simplify` statement in line 8, that provides the readable result $(a, b, c)^T$.

Example 10.5 In Example 5.3 in the SymPy chapter, we showed that the polynomials

$$p_1(x) :\equiv 1, \quad p_2(x) := x - 1/2, \quad p_3(x) := x^2 - x + 1/6$$

constitute a basis of the vector space $\mathbb{P}_2$ of polynomials of degree 2. In Maple we do it like this:

```
1 p[1] := 1: p[2] := x - 1/2: p[3] := x^2 - x + 1/6:
2 p_lin := add(c[i]*p[i], i = 1..3):
3 q := add(a[i]*x^i, i = 0..2):
4 r := q - p_lin:
5 solve({coeff(r, x, 0) = 0, coeff(r, x, 1) = 0,
6         coeff(r, x, 2) = 0}, {c[1], c[2], c[3]});
```

$$\{c_1 = 1/3\, a_2 + 1/2\, a_1 + a_0, \; c_2 = a_2 + a_1, \; c_3 = a_2\}$$

In line 1 we define the polynomials as term expressions. In line 2 we generate a linear combination of the p_i with placeholder coefficients c_i. Line 3 defines an arbitrary but fixed polynomial q of degree 2. In line 4 the *difference polynomial r* between q and the linear combination of the p_i is defined. In lines 5 and 6 the coefficients c_i are computed, such that the difference r evaluates to the zero polynomial. The output confirms that such c_i exist.

The Package `LinearAlgebra`

The `LinearAlgebra` package contains a number of advanced linear algebra methods. It is loaded with the command

```
with(LinearAlgebra);
```

In response, Maple displays a list of all functions in the package. As with other Maple commands, the output can be suppressed by deleting the semicolon, or replacing it with a colon.

For example, we can now compute the transpose of a matrix A with `Transpose(A)`. Other commands include `MatrixInverse(A)`, `LinearSolve(A,b)` or the construction of standard matrices like `IdentityMatrix(3)`.

The command `NullSpace(A)` is often useful, which determines the subspace of the vectors v with $Av = 0$, i.e. the kernel of the linear mapping induced by matrix multiplication.

Example 10.6 As in the SymPy Example 5.4 we show how to check that the vectors

```
u1 := <1, 0, 2>:   u2 := <0, 1, 1>:   u3 := <1, 2, -1>:
```

are linearly independent. For this we collect them in a matrix

```
A := <u1 | u2 | u3>;
```

$$A := \begin{bmatrix} 1 & 0 & 1 \\ 0 & 1 & 2 \\ 2 & 1 & -1 \end{bmatrix}$$

That A is regular can be confirmed by different observations, such as `Rank(A)` = 3, `Determinant(A)` = -5 or `NullSpace(A)` = $\varnothing$.

We determine the linear combination of the u_i to represent an arbitrary vector, say, $x = (8, 2, -4)$:

```
x := <8, 2, -4>
c := LinearSolve(A,x);
```

$$c := \begin{bmatrix} \frac{18}{5} \\ -\frac{34}{5} \\ \frac{22}{5} \end{bmatrix}$$

The test confirms the solution, the following computation returns the vector $(0, 0, 0)^T$:

```
c[1]*u1 + c[2]*u2 + c[3]*u3 - x;
```

Example 10.7 Finally we recast Example 5.5 from SymPy to get a matrix with a non-empty kernel:

```
C := Matrix([[1, 3, -1, 2], [0, 1, 4, 2],
             [2, 7, 2, 6], [1, 4, 3, 4]]):
NullSpace(C);
```

$$\left\{ \begin{bmatrix} 4 \\ -2 \\ 0 \\ 1 \end{bmatrix}, \begin{bmatrix} 13 \\ -4 \\ 1 \\ 0 \end{bmatrix} \right\}$$

Note that, like SymPy, Maple selects basis vectors with integer entries.

Eigenvectors and -values

For the computation of eigenvectors, the Maple package `LinerarAlgebra` provides the function `Eigenvectors`.

Example:

```
1 A := Matrix([[1, 2, -1], [4, 0, 1], [-7, -2, 3]]):
2 with(LinearAlgebra):
3 res := Eigenvectors(A);
```

$$res := \begin{bmatrix} 1 + \sqrt{15} \\ 1 - \sqrt{15} \\ 2 \end{bmatrix}, \begin{bmatrix} -\dfrac{(-1+\sqrt{15})\sqrt{15}}{15+7\sqrt{15}} & \dfrac{(-1-\sqrt{15})\sqrt{15}}{15-7\sqrt{15}} & 0 \\ -\dfrac{-15+4\sqrt{15}}{15+7\sqrt{15}} & -\dfrac{-15-4\sqrt{15}}{15-7\sqrt{15}} & 1/2 \\ 1 & 1 & 1 \end{bmatrix}$$

In the result `res`, the first component `res[1]` is a column vector that holds the eigenvalues, the second `V := res[2]` a matrix whose columns store the associated eigenvectors. The columns can be accessed by `Column(V,i)`.

For example, we do the test for the second eigenvalue:

```
4 e_vals :=res[1]:
5 V := res[2]:
6 A.Column(V,2) - e_vals[2]*Column(V,2):
7 simplify(%);
```

$$\begin{bmatrix} 0 \\ 0 \\ 0 \end{bmatrix}$$

Note that we have suppressed the output in line 6; again, it is only the `simplify` statement that renders the result recognizable.

$n \times n$-Matrices with $n \geq 5$

In the corresponding section in the SymPy chapter, we recalled that it is in general impossible to determine the eigenvalues of a large matrix exactly.

Here is Maple's attempt to compute the eigenvalues of the 5×5 variant of the Hilbert matrix with `Eigenvalues`:

```
H5 := Matrix(5, 5, (i,j) -> 1/(i+j-1));
Eigenvalues(H5);
```

As an output we get 5 lines according to the pattern

$$1/5\,RootOf(85349376\,\_Z^5 - 762725376\,\_Z^4 + 741667248\,\_Z^3 - 40915248\,\_Z^2$$
$$+61501\,\_Z - 1, index = 1)$$

which we do not discuss further. Let's try a numerical evaluation instead:

```
Eigenvalues(evalf(H5));
```

We then get 5 solutions of the form

$$1.56705069109486139 + 0.0\,i$$

with disappearing imaginary part. As already mentioned, one must expect complex solutions when determining eigenvalues. In symmetric matrices like the Hilbert matrix, however, all eigenvalues are real values in $\mathbb{R}$.

If we give Maple the additional information about the matrix, we finally come to an acceptable result:

```
unassign('H5'):
H5 := Matrix(5, 5, (i,j) -> 1/(i+j-1), shape=symmetric):
Eigenvalues(evalf(H5));
```

$$\begin{bmatrix} 0.0000032879392700686887 0 \\ 0.000305898025240842271 \\ 0.0114074915726152006 \\ 0.208534218668012500 \\ 1.56705069109486050 \end{bmatrix}$$

10.4 Calculus

Maple has a wealth of knowledge about the formation of limits, and offers a wide range of functions to deal with derivation and integration, both on a symbolic level and for numerical evaluations.

Derivation

As an introductory example, we compute the derivative f' of a function f as a limit of difference quotients, according to

$$f'(x) = \lim_{h \to 0} \frac{f(x+h) - f(x)}{h}.$$

We consider the function $f : x \mapsto ax^2$ with a symbolic parameter a.

Here is the Maple program:

```
1 f := x -> a*x^2
2 delta_f := (x,h) -> (f(x+h) - f(x))/h        # difference quotient
3 f_prime := x -> limit(delta_f(x,h), h = 0)   # limit
4 f_prime(3)
```

$$2ax$$

Unsurprisingly, Maple itself is quite good at derivatives. The Maple function `diff` returns the algebraic expression for the derivative of a function f at a point x. For the function `f` in line 1 above, for example, `diff(f(x),x)` again yields the term $2ax$.

The `diff` command knows all built-in functions and the usual derivation rules. For example, for `diff(x*sin(cos(x)),x)`, Maple returns the correct result

$$\sin(\cos(x)) - x \sin(x) \cos(\cos(x)).$$

This works also for higher-degree and partial derivatives:

```
1 w := (x,y) -> sin(y - a*x);   # example function
2 diff(w(x,y), x);
3 diff(diff(w(x,y), x), x);
4 diff(w(x,y), x, x);
5 diff(w(x,y), x$2, y$3);
```

Line 2 computes the first partial derivative $\dfrac{\partial}{\partial x} w$, line 3 the second $\dfrac{\partial^2}{\partial x^2} w$. Line 4 shows a direct specification of the second derivative. Line 5 yields $\dfrac{\partial^3}{\partial y^3} \dfrac{\partial^2}{\partial x^2} w$.

Note, however, that `diff` returns the algebraic *term expressions* of the derivatives, *not the derivative functions themselves*. But with the previously discussed `unapply` operator we can construct the desired *functions*.

It is also possible to generate the derivative functions directly with the *differential operator* D. The input `D(sin)` returns the actual cosine *function* cos, which can then also be applied as such:

```
D(sin)                # Out: cos
D(sin)(x) - cos(x)    # Out: 0
```

The result 0 confirms that `D(sin)` in fact coincides with the function cos.

The differential operator can also deal with higher and partial derivatives. For example, for the function `w:=(x,y) -> x^4*y^4 + x*y^2`, applying `D[1](w)` we get the function

$$\frac{\partial}{\partial x} w: (x, y) \mapsto 4x^3 y^4 + y^2.$$

Here the number 1 enclosed in brackets refers to the first argument of w. Other uses, such as `D[2](w)`, `D[2,2](w)` or `D[1$2,2$3](w)` should now be clear.

Integration

Maple also offers powerful options for symbolic integration. Definite and indefinite integrals are obtained with the command `int`, where, if no integration limits are specified, an antiderivative with integration constant 0 is computed. `int` knows the antiderivative functions of all built-in functions (if they exist) and an ever growing set of sophisticated integration rules.

For $\int \sin(x)^2\, dx$ we get

```
int(sin(x)^2, x);
```
$$-1/2\, \sin(x)\cos(x) + x/2$$

For the definite integral $\int_0^\pi \sin(x)^2\, dx$:

```
int(sin(x)^2, x = 0..Pi);
```
$$\pi/2$$

If Maple cannot find a solution, the system returns the expression unchanged:

```
int(exp(x^3), x = 0..2);
```
$$\int_0^2 e^{x^3}\, dx$$

Then there still remains the alternative to request a numerical evaluation:

```
evalf(%)   # Out: 276.8528882
```

Occasionally you need to provide additional information to get a useful solution. It is known that

$$\int_1^x 1/s\, ds = \ln(x) \quad \text{for all } x > 0.$$

If we want to verify this with Maple, we have to explicitly supply the restriction $x > 0$:

```
int(1/s, s = 1..x) assuming x > 0;
```
$$\ln(x)$$

Besides the `int` function there is also the `Int` variant (with a capital I), which is used when you want the integral representation without further transformation attempts.

Integral Norm

Consider the standard scalar product on the vector space $C[a, b]$ of continuous functions over an interval $[a, b]$, defined by

$$\langle f, g\rangle := \int_a^b (f \cdot g)(x)\, dx,$$

which, in turn, induces the following integral norm:

$$\|f\| := \langle f, f\rangle^{1/2}.$$

Example 10.8 We show that the functions sin and cos are orthogonal in the space $C[0, \pi]$ with respect to this scalar product:

```
int(sin(x)*cos(x), x = 0...Pi);    # Out: 0
```

For the norm of, say, sin we get $\| \sin \| = 1/2\sqrt{2\pi}$ with

```
sqrt(int(sin(x)*sin(x), x = 0...Pi));
```

Exercise Show that the basis of the $\mathbb{P}_2$ space considered in Example 10.5 is actually an *orthogonal* basis, when considered as a function space over the interval $[0, 1]$. How can this basis be converted to an ortho*normal* basis?

10.5 Interpolation with Spline Functions

An nth degree spline is a special function, defined piecewise by polynomials of degree n. At the points, where two polynomial pieces meet, certain conditions are set, for example, that the spline is $(n - 1)$ times continuously differentiable.

Originally, "spline" was a term for elastic rulers, that were bent to pass through a number of predefined points (also called *knots*). These were used to make technical drawings for shipbuilding and construction by hand.

For handling splines, Maple provides the function Spline.

Example 10.9 (Cubic Spline Interpolation) We interpolate 4 equidistant points in the interval $[0, 1]$ with cubic splines.

More precisely, we choose $x_0 = 0$, $x_1 = 1/3$, $x_2 = 2/3$, $x_3 = 1$, and set $y_0 = 0$, $y_1 = 1/3$, $y_2 = 1$, $y_3 = 3/4$ as associated y-values:

```
1 n := 3:
2 x_vec := Vector(n+1, i -> (i-1)/n):   # column vector size 4
3 y_vec := Vector([0, 1/3, 1, 3/4]):
4 pts := <x_vec | y_vec>:                # 4-by-2 matrix
```

Then we generate the interpolating spline function:

```
5 with(CurveFitting):                    # contains the Spline function
6 spl3 := Spline(pts, x);
```

$$\begin{cases} \dfrac{81\,x^3}{20} + \dfrac{11\,x}{20} & x<1/3 \\ -\dfrac{45\,x^3}{4} + \dfrac{153\,x^2}{10} - \dfrac{91\,x}{20} + \dfrac{17}{30} & x<2/3 \\ \dfrac{36\,x^3}{5} - \dfrac{108\,x^2}{5} + \dfrac{401\,x}{20} - \dfrac{49}{10} & otherwise \end{cases}$$

Observe that the spline is composed of polynomial pieces of degree 3. It is readily verified that function values as well as the derivatives (1st and 2nd) coincide at the knots between the subintervals.

Note that the spline expression is given as a *term*, which we can convert into a real function with unapply(spl3,x). Here we just want to plot it, however, and we can leave it as it is, in term form.

We plot the spline together with the points to be interpolated:

```
7  splplot := plot(spl3, x = 0..1):
8  with(plots):  # contains 'pointplot'
9  ptsplot := pointplot(pts, symbol=solidcircle,
10                      symbolsize=15, color=black):
11 display(splplot, ptsplot);
```

The result is shown on the left in Fig. 10.2 below, together with a *polygonal interpolation* explained in the following.

Polygonal Chains

By default, the `Spline` function produces polynomial chains of degree 3. If others are required, they must be specified explicitly. In various contexts we need *polygons*, i.e. splines of degree 1.

Example 10.10 (Polygonal Interpolation) We continue with the program in Example 10.9 and get:

```
12 spl1 := Spline(pts, x, degree=1);
```

$$spl1 := \begin{cases} x & x<1/3 \\ -1/3 + 2\,x & x<2/3 \\ 3/2 - 3/4\,x & otherwise \end{cases}$$

```
13 plot(spl1, x = 0..1);
```

The result is shown on the right in Fig. 10.2 below.

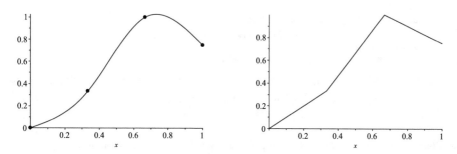

Fig. 10.2 On the left, cubic spline interpolation in Example 10.9. On the right, polygonal interpolation in Example 10.10

Example 10.11 (Hat Functions) We continue with the program in the examples above. Among others in the solution of differential equations according to the *finite element method*, often the following splines of degree 1 are used, the so-called *hat* or *tent functions* φ_i, generated like this:

```
14 plot_arr := Array(1..4):    # to store 4 plots in line 19
15 for i from 1 to 4 do
16        y_hat := Vector(n+1):  # 0 initialized
17        y_hat[i] := 1:         # peak re-defined
18        phi[i] := Spline(x_vec, y_hat, x, degree=1):
19        plot_arr[i] := plot(phi[i], x = 0..1):
20 end do:
21 display(plot_arr);
```

The graphs are shown in Fig. 10.3. These hat functions form a *basis* for all polygonal chains over the mesh points above.

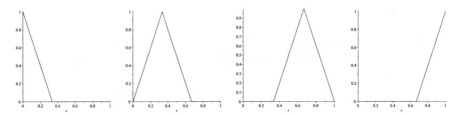

Fig. 10.3 Hat functions defined in Example 10.11

Example 10.12 (Hat Function Interpolation) To illustrate, let's continue with the above program again. In particular, we use the hat functions φ_i, implemented by the phi[i] in Example 10.11.

We consider a linear combination of the basis functions $f(x) := \sum_{i=1}^{4} c_i \varphi_i(x)$ and determine the coefficients such that $f(x_i) = y_i$ for the value pairs in Example 10.9:

```
22 f_expr := add(c[i]*phi[i], i = 1..4):
23 f := unapply(f_expr, x):
24 sols := solve({f(0) = 0, f(1/3) = 1/3, f(2/3) = 1, f(1) = 3/4},
25                {c[1], c[2], c[3], c[4]}):
26 assign(sols):
27 plot(f(x), x = 0..1);  # output exactly as in line 12 above
```

Note that we use the function f in line 22, and not the term f_expr. This is mainly due to writing economy. Instead of 'f(0) = 0' we could just as well have passed 'eval(f_expr,x=0) = 0' etc. as arguments to the solver.

Remark In fact, in the present simple case we do not need Maple to find the coefficients c_i. At point x_i, exactly one of the basis hat functions takes on a value $\neq 0$, namely $\varphi_i(x_i) = 1$. Therefore it is immediately clear that $c_i = y_i$.

The fact that the hat functions vanish outside off a small ("finite") vicinity around the evaluation points x_i makes them particularly suitable for the numerical computation of differential equations. The so-called *finite element method* is based on precisely this idea. We take a closer look at it below and then discuss the method in detail in the chapter on the FEniCS project.

10.6 Differential Equations

Maple offers many ways to deal with ordinary differential equations, both symbolically and numerically.

Solution by Direct Integration

Direct integration can often solve simple differential equations without detours:

Example 10.13 (Simple Boundary Value Problem) Consider the BVP

(1)
$$u''(x) = -(1 + x), \quad 0 < x < 1,$$
$$u(0) = 0, \ u(1) = 0.$$

It is clear that (1) can be solved by direct twofold integration. We have already noted that Maple does not add any integration constant, so we explicitly introduce c1 as such. So first we get a formula for u' by

```
1 int(-(1 + x), x) + c1;
```
$$-x - 1/2\,x^2 + c1$$

A second integration with the addition of another integration constant yields:

```
2 int(%,x) + c2;
```
$$-1/2\,x^2 - 1/6\,x^3 + c1\,x + c2$$

This expression is a *term*. Again it seems more convenient to work with a *function*, so we call unapply:

```
3 u := unapply(%,x);
```
$$x \mapsto -1/2\,x^2 - 1/6\,x^3 + c1\,x + c2$$

For the function u, we can now determine the constants $c1$ and $c2$:

```
4 sols := solve({u(0) = 0, u(1) = 0}, {c1, c2});
```
$$sols := \{c1 = 2/3, \ c2 = 0\}$$

Finally, we can *assign* the found values to the variables with

```
5 assign(sols):
6 u(x);
```
$$-1/6\,x^3 - 1/2\,x^2 + 2/3\,x$$

We can let Maple verify the result by computing diff(u(x),x$2) and the values u(0) and u(1).

Example 10.14 The same result is obtained with the built-in Maple function dsolve. All we have to do is to formulate the BVP and the boundary conditions in Maple, and then apply dsolve:

```
ode := diff(u(x), x, x) = -(1 + x):
bc := u(0) = 0, u(1) = 0:
dsolve({ode, bc});
```

The Euler Method as Maple Procedure

We define a Maple procedure myeuler to solve differential equations of the form

(1)
$$u'(x) = f(x, u), \quad 0 < x < 1,$$
$$u(0) = u_0$$

approximatly by the Euler method:

```
 1 myeuler := proc(f, u0, n)
 2     local dx, pts, i:
 3     dx := 1/n:
 4     pts := Matrix(n+1,2):  # to store (x,u)-pairs
 5     pts[1,1] := 0:         # start of x-intverval
 6     pts[1,2] := u0:        # initial u-value
 7     for i from 1 to n do
 8         pts[i+1,1] := pts[i,1] + dx:
 9         pts[i+1,2] := pts[i,2] + f(pts[i,1], pts[i,2])*dx: end do:
10     return pts:
11 end proc:
```

In line 1 'myeuler' is declared as a procedure with three arguments: the function f, the initial value u_0 from (1) and the number n to specify $n + 1$ evaluation points.

In line 2 three local variables are declared: dx for the length of the subintervals specified in line 3, the matrix pts to store the $(n + 1)$ value pairs and finally the variable i is simply the counter that is used in the for loop in line 7.

Beginning with initial values $(x, u) = (0, u_0)$, the subsequent values are computed according to the Euler method and appended to pts.

In line 10 the completed matrix pts is returned as a result of the procedure.

We check the algorithm for

(2)
$$f(x, u) := \frac{u(x)}{1 + x^2}, \quad u_0 := 1, \quad n = 10:$$

```
12 f := (x,u) -> u / (1 + x^2):
13 u_eul := myeuler(f, 1, 10):
```

For later use, we generate the representation of the points in the discrete solution:

```
14 with(plots):                    # contains the function pointplot
15 plot1 := pointplot(u_eul, symbolsize=15):
```

For comparison, we determine the exact solution with the integrated Maple function dsolve, for which the ODE and the initial condition are required as input arguments:

```
16 ode := diff(u(x), x) = u(x)/(1 + x^2):
17 ic := u(0) = 1:
18 sol := dsolve({ode, ic}):
```
$$sol := u(x) = e^{\arctan(x)}$$

We extract the right hand side $e^{\arctan(x)}$ from sol:

```
19 u := rhs(sol);
```

Finally, we plot the result together with the discrete approximation we established earlier:

```
20  plot2 := plot(u, x = 0..1):
21  display(plot1, plot2);
```

The result is shown in Fig. 10.4.

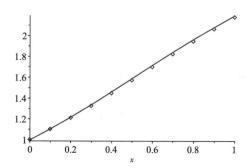

Fig. 10.4 Solution of Equation (1) with f defined in (2). Dots indicate discrete solution obtained with the Euler method, the solid line represents exact solution obtained with the built-in Maple function dsolve

10.7 Galerkin Method

In Sect. 5.6 in the SymPy chapter we illustrated the Galerkin method to solve differential equations. In the following, we show how to implement a corresponding solution program in Maple.

We once again consider the second order boundary value problem

(1)
$$-u'' = f \quad \text{over } [0, 1], \quad \text{where } f(x) := x\,(x + 3)\,e^x,$$
$$u(0) = u(1) = 0.$$

The Variational Problem

Recall that (1) is equivalent to the variational problem:

Determine u with $u(0) = u(1) = 0$ such that

(2)
$$-\int_0^1 u'' v = \int_0^1 f v$$

for every sufficiently smooth function $v: [0, 1] \to \mathbb{R}$ with $v(0) = v(1) = 0$.

Again we denote the left-hand side in (2) as $a(u, v)$ and the right-hand side as $L(v)$.

The Maple Program

We assume the same basic functions as in Sect. 5.6 in the SymPy chapter, in Maple expressed by

```
1 phi[1]:= x*(1 - x):
2 phi[2]:= x*(1/2 - x)*(1 - x):
3 phi[3]:= x*(1/3 - x)*(2/3 - x)*(1 - x):
4 phi[4]:= x*(1/4 - x)*(1/2 - x)*(3/4 - x)*(1 - x):
```

Next come the specification of the bilinear form a, the linear form L, the stiffness matrix A, and the load vector b:

```
5 a := (u,v) -> -int(diff(u,x,x)*v, x = 0..1):
6 f := x*(x + 3)*exp(x):
7 L := v -> int(f*v, x = 0..1):
8 A := Matrix(4, 4, (i,j) -> a(phi[j], phi[i])):
9 b := Vector(4, j -> L(phi[j])):
```

The matrix equation $Ac = b$:

```
10 with(LinearAlgebra):
11 c := LinearSolve(A,b):
```

Conversion of the solution to a function (term):

```
12 u := add(c[i]*phi[i], i = 1..4):
```

The solution can then be plotted with

```
13 plot(u, x = 0..1);
```

Exact Solution

Recalling that Maple does not add any integration constants, we introduce c_1, c_2 explicitly:

```
14 int(f,x) + c1:
15 int(%,x) + c2:
16 u_e := unapply(%,x):
17 sols := solve({u_e(0) = 0, u_e(1) = 0}, {c1, c2}):
18 assign(sols):
19 u_e(x);
```

$$-xe^x + x^2 e^x$$

As in the corresponding SymPy example, the deviation between the approximate and the exact solution is so small that it does not show up in the plot. As mentioned, in such cases, it is useful to plot the *difference* u-u_e. This is left to the reader.

10.8 Finite Element Method

In Sect. 10.4 we introduced the hat functions in the context of splines. If we choose the basic functions as suitable hat functions in the Galerkin approach, we come to the finite element method FEM. The FEM is discussed in detail later in the FEniCS chapter. Here we only illustrate the basic idea for the BVP (1) in Sect. 10.7 above.

We would like to proceed in the same way as in the general Galerkin case in the last section, based directly on the variational form

(1)
$$-\int_0^1 u''v = \int_0^1 fv.$$

However, here comes a crucial caveat: A trial function u expressed as $\sum c_i\varphi_i$ with hat functions φ_i has no usable second derivative. Hat functions can only be derived piecewise, and in particular the derivative functions φ_i' are *not continuous*.

Note that the derivatives *can* however be integrated and thus also u'. In fact, that is what we actually need.

We convert the left-hand side in (1) to an equivalent expression that only refers to first derivatives. This is obtained by partial integration, as follows:

(2)
$$-\int_0^1 u''v = \int_0^1 u'v' - u'v\Big|_0^1.$$

By the assumption $v(0) = v(1) = 0$, the second term on the right-hand side of (2) vanishes. Putting it all together, we get the following equivalent formulation for (1):

(3)
$$\int_0^1 u'v' = \int_0^1 fv.$$

The equation (3) can now be used as a basis for the Galerkin method. Again we refer to the left hand side as $a(u, v)$, and the right hand side as $L(v)$.

The Maple Program

We consider a sequence $0 = x_0 < x_1 < \cdots < x_{10} = 1$ of equidistant evaluation points and define the corresponding hat functions as in Sect. 10.4:

```
1  with(LinearAlgebra):
2  with(CurveFitting):
3  n := 10:
4  x_vec := Vector(n+1, i -> (i-1)/n):
5  for i from 1 to n+1 do
6      y_hat := Vector(n+1):
7      y_hat[i] := 1:
8      phi[i] := Spline(x_vec, y_hat, x, degree=1):
9  end do:
```

We come to the forms a and L from (3):

```
10  a := (u,v) -> int(diff(u,x)*diff(v,x), x = 0..1):
11  f := x^2:
12  L := v -> int(f*v, x = 0..1):
```

We turn to the stiffness matrix A. We could proceed as in the general Galerkin method and again create A directly with all entries of the form: `a(phi[i],phi[j]))`. But this does not take into account that A will now be a *sparse* matrix, more precisely a tridiagonal matrix. The reason is that by choosing the φ_i as local hat functions we get in particular

$$\varphi_i' \cdot \varphi_j' \equiv 0 \quad \text{for} \quad |i - j| > 1.$$

So we first create a sparse matrix with only zero entries and then populate it with the values $a(\varphi_i, \varphi_j) \neq 0$.

Moreover, the matrix is obviously symmetric, such that it is sufficient to specify the lower triangular part:

```
13  A := Matrix(n-1, n-1, shape=symmetric, storage=band[1, 0]):
```

Here `storage=band[1,0]` means that only the main diagonal and the first subdiagonal have to be stored.

The main diagonal in A has $(n - 1)$ entries:

```
14  for i from 1 to n-1 do A[i,i] := a(phi[i+1], phi[i+1]): end do:
```

Correspondingly for the subdiagonal:

```
15  for i from 1 to n-2 do A[i+1,i] := a(phi[i+1], phi[i+2]): end do:
```

Note that the border-adjacent hat functions φ_1 and φ_{11} are not used. The reason is simply that equation (1) is homogeneous, i.e., u assumes the value 0 at both boundary points, so that hat functions with function value $\neq 0$ at the boundaries could not contribute to the solution anyway.

We come to the load vector b, where we again only need to consider the inner hat functions:

```
16  b := Vector(n-1, j -> L(phi[j+1])):
```

We solve the equation system $Ac = b$ to find the coefficient values c_i for the approximate solution $u = \sum c_i \varphi_i$ at the inner evaluation points:

```
17  c := LinearSolve(A,b):
```

The solution vector c is then extended to include the coefficients $c_1 = c_{11} = 0$ for φ_1 and φ_{11}:

```
18  c := Vector(n+1, [0, c, 0]):
```

Finally the solution function u is generated and plotted, as shown in Fig. 10.5.

```
19  u := add(c[i]*phi[i], i = 1..11):
20  plot(u, x = 0..1);
```

Remark As pointed out in Remark 10.5, the hat functions assume the value 1 at their peak point by definition , so there is, strictly speaking, no need to formulate the solution u as the sum $u = \sum c_i \varphi_i$. From the c_i values themselves we could generate u in line 19 directly as

```
u := Spline(x_vec, c, x, degree=1):
```

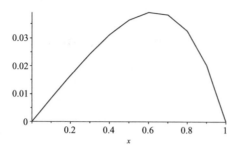

Fig. 10.5 Finite-element approximation of BVP (1) in Sect. 10.7

10.9 Exercises

Ordinary Differential Equations

Exercise 10.1 Use the Maple function dsolve to solve the BVP

$$u''(x) = u(x),$$
$$u(0) = 0, \ u'(1) = -1,$$

already discussed in Exercise 5.7 in the SymPy chapter, and plot the solution.

Hints (1) Use D(u)(1)=-1 to represent the second boundary condition. (2) The function to be plotted is returned as the right-hand side of the dsolve result.

Exercise 10.2 Show that Maple finds no symbolic solution for the equation

$$u''(x) + \sin u(x) = 0,$$
$$u(0) = 0, \ u'(0) = 1.$$

Use dsolve with the option 'type=numeric' to retrieve a numerical solution. Plot it with the function odeplot from the package plots. Note that odeplot works directly with the dsolve result, without the need to extract the right-hand side.

Polynomial Approximation and Interpolation

The Weierstrass approximation theorem states that every continuous function defined on a closed interval $[a, b]$ can be uniformly approximated as closely as desired by a polynomial function: For every function $f \in C[a, b]$ there is a sequence P_n of polynomials, such that

$$\lim_{n \to \infty} \left(\max_{a \le x \le b} |f(x) - P_n(x)| \right) = 0.$$

At first glance, it may seem intuitively plausible to choose the P_n as interpolation polynomials of degree n on a sequence of increasingly dense sets of equidistant evaluation points. But that does not lead to success; on the contrary:

Exercise 10.3 Consider the *Runge function*

$$f(x) = \frac{1}{1 + 25x^2}, \quad x \in [-1, 1].$$

(1) Use `PolynomialInterpolation` from the package `CurveFitting` to compute the interpolation polynomials P_n of degree n on the $n+1$ equidistant evaluation points

$$x_i := -1 + (i - 1)\frac{2}{n}, \quad i = 1, \ldots, n + 1,$$

for $n = 5$, $n = 11$ and $n = 17$. Draw the graph of f and the P_n.

(2) Make a conjecture as to how the approximation evolves for increasing n. Support your conjecture by sketching the graph of the function

$$\epsilon_n(x) := f(x) - P_n(x)$$

for the n-values above.

Exercise 10.4 Repeat the same reasoning as in the last exercise, with the difference that now you consider the following *Chebyshev nodes* instead of the equidistant interpolation points:

$$x_i := -\cos\left(\frac{i - 1}{n}\pi\right), \quad i = 1, \ldots, n + 1.$$

Remark In a sense, the Chebyshev nodes x_i are also based on an equidistant division. They result from the projection of a sequence of points $p_i = (x_i, y_i)$ on the upper unit semicircle, equidistantly distributed on the arc.

Interpolation with Legendre Polynomials

In the exercises for the SymPy chapter we defined the Legendre polynomials and showed that they are pairwise orthogonal in the space $C[-1, 1]$ with respect to the integral scalar product. Here we discuss how they can be used in polynomial interpolation.

To recall: The standard integral scalar product in $C[-1, 1]$ is defined by

$$\langle f, g \rangle := \int_{-1}^{1} f(x)g(x)\,dx.$$

Beginning with the *monomials* $M_n(x) := x^n$, the Legendre polynomials L_n are then inductively defined as

$$L_0 := M_0, \quad L_n := M_n - \sum_{i=0}^{n-1} \frac{\langle M_n, L_i \rangle}{\langle L_i, L_i \rangle} L_i.$$

To approximate a given function $f \in C[-1, 1]$, we can then use the polynomials P_n given by

$$P_n := \sum_{i=0}^{n} \frac{\langle f, L_i \rangle}{\langle L_i, L_i \rangle} L_i.$$

In fact, these P_n are the optimal approximation polynomials of degree n for f, in the sense of

$$\text{minimize} \quad \|f(x) - P(x)\| \quad \text{over all polynomials } P \text{ of degree } n,$$

where $\| \cdot \|$ denotes the integral norm, induced by the scalar product $\langle \cdot, \cdot \rangle^{1/2}$.

Exercise 10.5 To illustrate this, let's return to the Runge function f defined above.

(1) Approximate f with some of the P_n, and plot the result.

(2) Compute the sequence $\epsilon_n := \|f - P_n\|$ and test whether it confirms the convergence assumption.

Spline Interpolation

Another way to avoid the "boundary oscillation" problem of polynomial interpolation is to join polynomials of small degree to form polynomial chains. This was briefly discussed in Exercise 8.7 in the Julia chapter.

Exercise 10.6 Without using the package `CurveFitting`, define a cubic spline to interpolate the points (1,1), (2,4), (3,3).

Observe that in the resulting equation system, there are two degrees of freedom left. Now consider the spline generated by the Maple function `Spline` and explain how these degrees of freedom are used there.

Exercise 10.7 Approximate the Runge function by cubic splines on sets of equidistant evaluation points.

Part IV
Distributed Computing

Chapter 11
A Python Approach to Message Passing

Modern high-performance computing owes its success not least to the concurrent processing of complex computing tasks on distributed systems, in networks that consist of nodes of separate machines, or on individual machines with an architecture with multiple processors. In fact, most computers today have more than one processor core, so we can use distributed processing on standard laptops, not just supercomputers like before.

We can distinguish two main approaches to distributed computing, firstly *message passing* systems. This corresponds to a distributed group of cooperative workers who perform a common task within a network of nodes and send and receive messages via *shared channels*. On the other hand, the idea is to organize the communication between independent actors through concurrent access to a local *shared memory*.

Distributed computing and the theory of distributed systems in general, is still a field of intense research in computer science. Seminal contributions to the theoretical foundations were developed in the Institute for Instrumental Mathematics at the University of Bonn, Germany, in the late 1950s and later the GMD (German research center for information technology) by the then head of the university computing center, Carl Adam Petri. The author takes the opportunity to recommend his biography [13], where much of the background is illustrated.

In this chapter we develop the basics of message passing within the particularly user friendly language Python.

All examples are provided in a form that can be tested on normal home computers.

11.1 Introduction to the Message Passing Interface

A standard specification for message-passing computation is provided and maintained by the forum for the *Message Passing Interface* (MPI), which began its work in the early 1990s.

© Springer Nature Switzerland AG 2020
E. Smith, *Introduction to the Tools of Scientific Computing*, Texts in Computational Science and Engineering 25, https://doi.org/10.1007/978-3-030-60808-8_11

Originally developed for the programming languages C, C++ and the venerable Fortran, the basics of MPI have in the meantime been adapted also to Python.

Here we follow that latter route, more specifically, we discuss the mpi4py implementation developed by Lisandro Dalcin from Argentina.

For background and further reading, the tutorials mpi4py.readthedocs.io on the official website can be recommended. Another good source can be found on materials.jeremybejarano.com/MPIwithPython.

Installation

We assume the Python distribution *Anaconda* to be installed. Then the necessary MPI components can be downloaded and installed with the Terminal command

```
$ conda install mpi4py
```

Testing

To test the installation, we consider a file hello.py that contains only the single line print("Hello World"). Assuming that the file is stored in the user's home directory, where the Python interpreter finds it directly, we can run the program in the Terminal window as follows:

```
$ python hello.py
Hello World
```

Now, what is new, is that in our just installed MPI system, we can run *multiple* copies (here 2) of the program in parallel, using the command mpiexec with the '-n 2' option:

```
$ mpiexec -n 2 python hello.py
Hello World
Hello World
```

In fact, what happens is that mpiexec launches *two independent instances* of the Python interpreter, each of which then executes the program in hello.py and returns the output to the common console window.

Remark But note that we do not require the system to spawn the execution to different processors. The whole development can be illustrated, and will also work, on a single processor machine. Of course, the actual performance gain in parallel computing can only be obtained with multi-processor systems. As mentioned, today this holds for most modern computers.

However, on many systems the number n of instances is by default limited to the number of available computing nodes. To run, say, 8 instances on the 6 core system used here, requires to "*oversubscribe*" that number with e.g.

```
$ mpiexec -oversubscribe -n 8  python hello.py
```

Unfortunately, the comfortable programming environment Spyder, which we used in the previous Python chapters, only allows one interpreter instance to be executed at a time.

In the following, all our MPI programs are run from the Terminal command line in the form

```
$ mpiexec -n <number of parallel processes>  python <file>.py
```

The MPI Module

The main idea is that multiple processes should be enabled to *interact*. The two Python processes above know nothing of each other.

That changes if we tell each individual process that it is part of a larger whole.

For this (and a lot more) we need the Python package mpi4py, and there in particular the class-object MPI.COMM_WORLD, that defines a context of inter-process communication. It contains the essential attributes and methods we will use in the following, for instance the variable 'size' that stores the number of participating processes in our communication world, and a number rank for each, to distinguish the individual processes.

Example 11.1 We assume the program to be stored in a file hello_from.py:

```
1 from mpi4py import MPI
2 comm = MPI.COMM_WORLD
3 print("Hello World from process", comm.rank, "of", comm.size)
```

In line 1 we import the basic module MPI. The communication context is controlled by the object MPI.COMM_WORLD, which contains the variables size and rank and most of the methods to be introduced later.

In line 2 we introduce the common abbreviation comm for the communicator object MPI.COMM_WORLD, and in line 3 use it to represent the number of processes comm.size in the communication system and to assign a specific index number comm.rank from 0 to comm.size-1 to each.

Running the program hello_from.py for n = 2 processes we get

```
$ mpiexec -n 2 python hello_from.py
Hello World from process 1 of 2
Hello World from process 0 of 2
```

Note that the rank numbers do not induce any execution order. In fact, the order can change between different program runs.

11.2 Communicating Processes

Now it is time to establish communication between processes. Here send and recv are the most basic – and most important – operations.

Example 11.2 We assume the following code to be stored in a file hello_to.py:

```
1 from mpi4py import MPI
2 comm = MPI.COMM_WORLD
3 if comm.rank == 0:
4     comm.send("Hello World", dest=1)
5 if comm.rank == 1:
6     msg = comm.recv(source=0)
7     print("Message received:", msg)
```

Observe that the program contains separate code parts. Lines 1 and 2 are executed by all processes, line 4 only in process 0, and the block of lines 6 and 7 in process 1. Which block refers to which process is controlled by the if-conditions in lines 3 and 5. This is the standard way to schedule different process behavior in MPI systems.

The operation send in line 4 takes two arguments, the data to be sent and the destination to where it is sent.

In line 6 process 1 waits for data from process 0, and when it arrives, recv stores it in the variable msg, which is then printed to the screen in line 7.

A run for $n = 2$ processes yields

```
$ mpiexec -n 2 python hello_to.py
Message received: Hello World
```

Deadlocks

It is obvious that a receive operation must wait for the associated send operation. This strict synchronization order between send and receive is however a permanent source of run-time errors in parallel programming. Assume the following program to be stored in a file dl.py:

```
1 from mpi4py import MPI
2 comm = MPI.COMM_WORLD
3 msg = comm.recv(source = comm.size - 1 - comm.rank)
4 comm.send(22, dest = comm.size - 1 - comm.rank)
5 print(msg)
```

We try to run the code for two processes:

```
$ mpiexec -n 2 python dl.py
```

Then in line 3 each of the processes mutually waits for the other one to send some data before it can continue to send its own message in line 4. The program hangs, it must be forced to abort, with the key combination "Ctrl-C".

Such a situation is called *deadlock*.

If lines 3 and 4 are exchanged, the program can continue to line 5 and print the message 22.

Notation Convention

In the examples above we illustrated the usual convention to write comm as an abbreviation for MPI.COMM_WORLD. In the following we opt for an even shorter notation.

We print MPI.COMM_WORLD variables and methods like **rank**, **size**, **send** and **recv** in bold to indicate that they have been provided according to the pattern

```
rank = MPI.COMM_WORLD.rank
```

Moreover, we will always assume that the module MPI is loaded in the form

```
from mpi4py import MPI
```

Summation

As a first "real" application, we consider an MPI program to compute the sum $\sum_{i=1}^{n} i$.

Example 11.3 The program begins with a preamble, importing the MPI module and declaring the variable and method abbreviations.

We then fix the upper limit n as

```
1  n = 100
```

We assume that the number size of processes evenly divides the number n of summands. If not, the program should abort with an error message:

```
2  assert n % size == 0  # prog. aborts if condition not satisfied
```

We partition the summation range $1, \ldots, n$ into size many contiguous segments of equal length:

```
3  k = rank * n//size + 1
4  l = (rank+1) * n//size + 1
```

Each process locally computes the sum over its own segment:

```
5  sm_loc = sum(range(k,l))
```

To illustrate, assume size = 4. For the process with, say rank = 2, we then get $k = 51$, $l = 76$. Recalling that in Python the upper bound l is not included in the range, in line 5 the numbers $51, 52, \ldots, 75$ are added up by the Python function sum and stored as the contribution from process 2 to the final result.

The root process 0 collects the partial sums from the other processes and adds them to its own contribution:

```
6   if rank != 0: send(sm_loc, dest=0)  # send from the other processes
7   else:                                # in root
8       sm = sm_loc                      # computed by root process itself
9       for i in range(1, size):
10          sm += recv(source=i)         # add received contributions
```

Finally the result is printed to the screen:

```
11  if rank == 0: print(sm)              # printing controlled by root
```

In line 11 we follow the good convention to print results only from the root process. During program development it may however be useful to insert print commands at arbitrary neuralgic points.

Assume the program to be stored in a file sum.py. It can then be distributed to any number of processes that evenly divides the internal variable $n = 100$, e.g.

```
$ mpiexec -n 4 python sum.py
5050
```

Point-to-Point vs. Collective Communication, the reduce and bcast Commands

Both send and recv operate on a point-to-point basis: send sends a message to one specific destination, recv is prepared to receive a message only from a specified source. However, MPI additionally provides various *collective communication* operators, with which one process can e.g. *collect* messages from a group of processes or *broadcast* a message to all processes.

The reduce Command

Here is a first example of a collective operation:

Example 11.4 In our sum program in Example 11.3 we collected the partial results through a series of coupled point-to-point send/receive operations.

Using the sophisticated reduce method, we can actually replace lines 6–10 above by a single line

```
sm = reduce(sm_loc)
```

What happens is that the sm_loc values from all processes are collected, then "reduced" to a representation as a single value, namely their sum, and then assigned to the variable sm in the *root* process.

In the other processes, the reduce command assigns a special Python value None to sm. This can be verified by deleting the 'if rank == 0:' clause to the print(sm) command in line 11.

Actually, reduce is even much more flexible. The computation of the sum is only the default option, with the meaning 'reduce(., op=MPI.SUM)'. Several others are available, like MPI.PROD or MPI.MAX with their obvious meanings.

The bcast Command

In the summation example, the same computation of 'n//size' is repeated in each process. Actually, it is sufficient to compute the quotient once in process 0, and then distribute the result to all. This can be achieved with the bcast command.

Example 11.5 To illustrate the idea, we replace lines 1–4 in Example 11.3 above with the following code:

```
1 q_loc = None
2 if rank == 0:
3     n = 100
4     assert n % size == 0
5     q_loc = n//size
6 q = bcast(q_loc)
7 k =    rank*q + 1
8 l = (rank + 1)*q + 1
```

In line 5 the quotient is computed in process 0 and broadcasted to each process' local variable q in line 6. Note that the command is executed in each process, hence the variable q_loc must be declared everywhere. In Python, declaration without initialization is not possible, hence q_loc must be assigned some (arbitrary) value. This happens in line 1.

Note that it is perfectly possible, and in fact very common, to use the same name for both variables q and q_loc.

Load Balancing

In our summation example we assumed that the number of processes evenly divided the number of summands, such that the work load assigned to the processes was equal. If that is not the case, then some measures must be taken.

A common way to deal with the general case is to distribute the remaining entries one by one to the first processes.

Example 11.6 Assume we wish to distribute the computation of $1 + 2 + \cdots + 100$ to 8 processes. Then the sum up to 96 can be treated as before. For the rest we let each of the processes $i \leq 3$ add $97 + i$ to their sm_loc-contribution.

Here is the complete program, in the comments illustrated for $n = 100$ and 8 processes:

```
1 tup = None                      # var needed for broadcast
2 if rank == 0:
3     n = 100
4     q = n//size; u = q*size; r = n % size
5     tup = q, u, r               # = 12, 96, 4
6 q, u, r = bcast(tup)
7 k =    rank*q + 1
8 l = (rank + 1)*q + 1
9 sm = sum(range(k,l))
10 if rank < r: sm += u+1   + rank # adds 97+i for ranks i = 0,1,2,3
11 sm = reduce(sm)
12 if rank == 0: print(sm)
```

11.3 Integral Approximation

Numerical integration is a good "real-world" candidate for distributed computing. In this section we consider integral approximation by the *midpoint* and *trapezoidal* rule.

The Midpoint Rule

Let $f: [a, b] \to \mathbb{R}$. To approximate the integral of f we divide the interval $[a, b]$ into n subintervals of equal length h, and sum up the areas $hf(x_i)$ where x_i is the midpoint of subinterval number $i = 0, 1, \ldots, n - 1$.

Formally:

$$\int_a^b f(x)\, dx \approx h \sum_{i=0}^{n-1} f\left(a + h(i + \frac{1}{2})\right), \quad \text{where } h = \frac{b - a}{n}.$$

In our parallel program we want to distribute the summation to the involved processes. Note that the summands are independent, hence the partition between them is arbitrary. For a given number n of subintervals and s processes we choose to let process number $r < s$ take care of the indices $i < n$ of the form $i = r + ms$, $m = 0, 1, \ldots$.

For instance for $n = 10$ and $s = 4$, this means that process 0 is responsible for three indices $i = 0, 4, 8$, process 1 for $i = 1, 5, 9$, process 2 for $i = 2, 6$, and process 3 for the remaining $i = 3, 7$.

Note that this distribution of the subintervals is not contiguous. The advantage of the present approach is that load balancing takes place automatically in the manner described above and entries that remain after equal partitioning, are assigned one by one to the first processes.

Example 11.7 Assuming the usual preamble is already loaded, here is the MPI-program:

We first fix a test function

```
1 f = lambda x:   4/(1 + x*x)
```

to be integrated over the unit interval $[0, 1]$.

The parameters needed in the local computations are set in the root process, and then broadcast to all participating processes:

```
2 tup = None
3 if rank == 0:
4     n = 100_000
5     a = 0.0; b = 1.0; h = (b - a)/n
6     tup = n, a, h
7 n, a, h = bcast(tup)
```

Then each process computes its contribution according to the pattern explained above:

```
 8  add = 0.0
 9  for i in range(rank, n, size):
10      x = a + h*(i + 0.5)
11      add += f(x)
12  int_loc = h*add
```

The local contributions are summed up and the result printed by the root process:

```
13  int = reduce(int_loc)
14  if rank == 0: print(int)
```

Distributed to 8 processes, the program returns the value 3.141592653598135, which is of course no coincidence, since in fact

$$\pi = \int_0^1 \frac{4}{1 + x^2}\, dx.$$

The Trapezoidal Rule

The trapezoidal rule is a more efficient approximation technique for the definite integral $\int_a^b f(x)\, dx$.

Assume a grid of the interval $[a, b]$ given by $n + 1$ equidistant points

$$x_i := a + ih \quad \text{with} \quad h = \frac{b - a}{n} \quad \text{for} \quad i = 0, \dots, n.$$

Let

$$I(k, l) := \frac{h}{2}\left(f(x_k) + 2 \sum_{i=k+1}^{l-1} f(x_i) + f(x_l) \right), \quad 0 \le k < l \le n,$$

where the sum term disappears if $k + 1 = l$.

Then by the trapezoidal rule:

$$\int_a^b f(x)\, dx \approx I(0, n).$$

Now, the crucial observation is that the evaluation of I can the distributed to parallel processes, since always

$$I(k, l) = I(k, p) + I(p, l), \quad k < p < l.$$

The MPI Program

We write a corresponding MPI program. In the midpoint approach, it was easy to take care of load balance. To use the same non-contiguous distribution here would however lead to a large overhead, since border parts would have to be calculated twice.

In the following program, for simplicity we assume the number of processes to evenly divide the number of subintervals.

As usual the program begins with the preamble, importing the MPI module and declaring the required variable and method abbreviations.

We then define an example function f, an interval $[a, b]$ over which to integrate and the number of equidistant grid points.

We use the same test function as before:

```
1 def f(x): return 4/(1 + x*x)
```

The parameters are prepared and broadcast:

```
2  tup = None
3  if rank == 0:
4      a = 0; b = 1; n = 100_000
5      h = (b - a)/n
6      from numpy import linspace
7      x = linspace(a, b, n+1)
8      q = n//size
9      tup = q, x, h
10 q, x, h = bcast(tup)
```

Each process determines its segment and computes the partial integral:

```
11 k = rank*q; l = k + q
12 int_loc = h/2*(f(x[k]) + 2*sum(f(x[k+1: l])) + f(x[l]))
```

Note that here 'sum' again denotes the standard Python function to add up elements in a collection.

As usual the partial results are summed up by root and then printed:

```
13 int = reduce(int_loc)
14 if rank == 0: print(int)   # printing controlled by root
```

Distributed to 8 processes the program returns the result 3.141592653573129.

11.4 Vector Dot Product

We first consider the special case that all vector segments can be evenly divided among the processing units. We then turn to the general case, and show how to to distribute segments in an economic manner.

Dot Product Evenly Divided

We develop an MPI program for the distributed scalar (or dot) multiplication of vectors, where the length of the involved vectors are evenly divisible by the number of processes.

More complex communication in mpi4py is based on *NumPy arrays*, so we import the needed components:

```
1 from numpy import array, zeros
```

The root process initializes two example vectors v, w and prepares for the distribution to all processes.

As an example we take $v = (1, 2, \dots, n - 1, n)$ and $w = (n, n - 1, \dots, 2, 1)$. Note that for the communication to go through, we have to declare them as NumPy arrays with float entries:

```
2 v = w = n_loc = None        # vars must be known in all processes
3 if rank == 0:
4     n = 12                  # number of processes must divide n ...
5     assert n % size == 0    # ... if not: abort
6     v = array([float(i) for i in range(1, n+1)])
7     w = array([float(i) for i in range(n, 0, -1)])
8     n_loc = n//size         # division with remainder 0
```

The number n_loc is broadcast to all processes:

```
9 n_loc = bcast(n_loc)
```

Each process prepares (initially empty) buffers of that size n_loc for the storage of their segment of the vectors v and w:

```
10 v_loc = zeros(n_loc); w_loc = zeros(n_loc)
```

The operator Scatter divides both vectors into size many pieces of equal length n_loc, and stores the subvectors in the individual buffers v_loc and w_loc:

```
11 Scatter(v, v_loc); Scatter(w, w_loc)  # capital letter S, see below
```

The local parts are multiplied:

```
12 dot_loc = v_loc @ w_loc  # recall: @ denotes scalar multiplication
```

and the partial results then "reduced" to the global result dot in the root process:

```
13 dot = reduce(dot_loc)
```

and finally printed:

```
14 if rank == 0: print(dot)
```

Interlude: Message Passing with pickle and NumPy array

In mpi4py, basic message passing uses methods from the pickle module, discussed in Sect. 3.8 in the Python chapter, more precisely the functions dump and load. This allows for high-level and very convenient communication.

For large data sets, however, this approach can be slow. In particular, for the communication of arrays, more fine-tuned operations may be more efficient.

The operators can generally be distinguished by their initial letter. pickle communication of Python objects uses operators with lower case initial letters like send

and recv, whereas the corresponding array-based operators are written with initial
capital, like Send and Recv.

In the example above, we used the *array*-based function Scatter to distribute the
vectors.

Dot Product with Load Balance

It is of course unsatisfactory, that we can as yet only input vectors, where the length n
is an integer multiple of the number p of processes.

As mentioned, a common way to deal with the general case, is to distribute the
remaining entries one by one to the first processes, such that, say, for $n = 10$, $p = 4$,
processes 0 and 1 receive 3 entries each, and the remaining processes 2 and 3 only 2.
Note that this is exactly the distribution in the integral approximation in Sect. 11.3.

We begin with the program right away:

```
1 from numpy import array, zeros
```

The root process defines example vectors as before and prepares for a balanced distribution:

```
2 v = w = sendcts = None
3 if rank == 0:
4     n = 10
5     v = array([float(i) for i in range(1, n+1)])
6     w = array([float(i) for i in range(n, 0, -1)])
7     n_loc = n//size  # even divisibility not required
8     sendcts = [n_loc for i in range(size)]
9     for i in range(n % size): sendcts[i] += 1
```

In the general case, the number n_loc cannot be assumed to uniformly describe the
size of the vector segments. Instead, in line 8 we collect the individual sizes in an
integer list sendcts, at first initialized with the entry n_loc for all processes.

With p processes there are still $r := n \bmod p$ entries to be distributed. In line 9
this is prepared for, increasing the segment size by 1 for each process $0, 1 \ldots, r - 1$.

The list sendcts, which now contains the segment sizes for all process, is broadcast:

```
10 sendcts = bcast(sendcts)
```

Each process 'rank' then provides local buffers of suitable size for the storage of its
subvectors:

```
11 v_loc = zeros(sendcts[rank]); w_loc = zeros(sendcts[rank])
```

In the previous example we used Scatter to distribute the vectors. Here the matter is
more complicated, since the splitting points are not determined by a constant offset
number n_loc, but rather by the entries in the list sendcts.

This time we have to use a generalized scatter function Scatterv, which however as
first argument requires significantly more information than Scatter:

```
12 Scatterv([v, sendcts, MPI.DOUBLE], v_loc)
13 Scatterv([w, sendcts, MPI.DOUBLE], w_loc)
```

The last lines can then be copied from the previous example:

```
14 dot_loc = v_loc @ w_loc
15 dot = reduce(dot_loc)
16 if rank == 0: print(dot)   # Out: 220.0
```

11.5 Laplace Equations

In all examples so far, communication between processes was limited to the initial distribution of data, and then the final recombining of the partial results.

In general however, also interleaved communication *during* the computation will be required. As an example, we discuss the solution of *partial differential Laplace equations* by the *method of finite differences*. For background we refer to Sect. 4.8 in the SciPy chapter.

Recall that a Laplace equation is a special Poisson equation $\Delta u = f$, where $f \equiv 0$. As an example we consider the following PDE:

Determine a function $u \colon \Omega \to \mathbb{R}$, $\Omega := [0, 1] \times [0, 1]$, such that

$$
\begin{aligned}
\Delta u &= 0 \quad \text{in } \Omega, \\
u &= g \quad \text{on the boundary } \partial\Omega
\end{aligned}
\tag{1}
$$

for a given function g, where we here assume the special case

$$
g(x, y) := \begin{cases} x(1 - x) & \text{if } y = 0, \\ 0 & \text{else.} \end{cases}
\tag{2}
$$

For our numerical solution we once again need a discretization corresponding to the stencil equation (4) in Sect. 4.8 in the SciPy chapter. Since in the present case the right-hand side is 0, after some manipulations we get the equivalent formulation

$$
u_{ij} = \frac{1}{4}\left(u_{i-1,j} + u_{i,j-1} + u_{i+1,j} + u_{i,j+1}\right), \quad i, j = 1, \dots, n.
\tag{3}
$$

Now the crucial observation is that we can interpret (3) as a *fix-point solution* of the iteration sequence

$$
u_{ij}^{(k+1)} := \frac{1}{4}\left(u_{i-1,j}^{(k)} + u_{i,j-1}^{(k)} + u_{i+1,j}^{(k)} + u_{i,j+1}^{(k)}\right), \quad i, j = 1, \dots, n, \quad k \geq 0.
\tag{4}
$$

The discretization of the boundary condition yields the start values for the iteration:

$$
u_{ij}^{(0)} := \begin{cases} x_j(1 - x_j) & \text{for } x_j := \frac{j}{n+1}, \quad \text{if } i = 0, \\ 0 & \text{else,} \end{cases} \quad i, j = 0, \dots, n+1.
\tag{5}
$$

A Distributed Iteration

We show how to distribute the iteration over *two* processes. We assume n to be an even number.

The task is divided *vertically* at the x-index $m = \frac{n}{2}$, such that one process P_{left} performs the computations of $u_{ij}^{(k+1)}$ for $j \le m$, and P_{right} the ones for $j \ge m + 1$.

We follow the iteration step "$k \to k + 1$" for P_{left}. The required values from step k are provided by P_{left} itself, *except* the ones for the "right-boundary" cases $u_{i,m}^{(k+1)}$, which also need the values $u_{i,m+1}^{(k)}$. These are provided by P_{right}.

Conversely, P_{right} needs P_{left} to provide the $u_{i,m}^{(k)}$ in order to compute the $u_{i,m+1}^{(k+1)}$.

In MPI jargon, the extra boundary points provided by the neighbor are referred to as "ghost points".

The MPI Program

As usual the program begins with our mpi4py preamble.

The root process 0 then prepares the matrix u to store the grid values u_{ij}:

```
1 if rank == 0:
2     from numpy import linspace, zeros    # needed only in root
3     n = 10; m = n//2                      # n evenly divided by 2
4     u = zeros((n,n))                      # solution matrix 0 initialized
5     x = linspace(0, 1, n)
6     u[0,:] = x*(1-x)                      # inclusion of boundary values
```

The matrix u is then split vertically, and the right half sent to the helper process 1:

```
7     ul = u[:, :m]                         # u evenly divided in left
8     ur = u[:, m:]                         # ... and right part
9     send(ur, dest=1)                      # ... sent to helper process
10 if rank == 1: ur = recv(source=0)        # ... and received
```

The Python function below is used to compute the iterative steps for the inner points, according to the sequence definition (4) above:

```
11 def stencil_it(u):
12     u_it = u.copy()                      # copy to avoid side effects
13     r, c = u.shape                       # number of rows and columns
14     for i in range(1, r-1):              # inner rows
15         for j in range(1, c-1):          # inner columns
16             u_it[i,j] = (u[i+1,j] + u[i-1,j]
17                         + u[i,j+1] + u[i,j-1]) / 4
18     return u_it
```

We come to the central part of the program, the iterative approximation, distributed between root and helper process. Note that in addition to the organizational preliminaries above, the root process 0 also assumes the left-hand part of the computation.

We need two more NumPy functions to add and delete the columns of ghost points required in the computation:

```
19  from numpy import column_stack, delete  # needed in both processes
```

We fix the number of iterations and start the loop. Note that the loop body consists of lines 22 up to 33.

```
20  iter = 40
21  for i in range(iter):                   # central iteration loop
```

In lines 22–27 we focus on the root process:

```
22      if rank == 0:
```

Root sends the current ghost points in the last column of ul to the helper, where it is received in line 30:

```
23          send(ul[:, -1], dest=1)         # '-1' denotes last column
```

Conversely, root receives the column of ghost points, sent by the helper in line 29, and appends it to its own matrix part:

```
24          v = recv(source=1)
25          ul_v = column_stack([ul, v])    # append v as right boundary
```

Now the inner grid points can be computed for the next iteration, and the ghost points then discarded:

```
26          ul_v = stencil_it(ul_v)         # computed new inner values
27          ul = delete(ul_v, -1, 1)        # delete v
```

The second argument '-1' in the delete function stands for 'last', the third '1' for 'column'.

We shift focus to the helper process. The procedure is essentially the same as in root, except that it is mirrored:

```
28      if rank == 1:
29          send(ur[:, 0], dest=0)          # sends leftmost column
30          v = recv(source=0)
31          v_ur = column_stack([v, ur])    # insert v as left boundary
32          v_ur = stencil_it(v_ur)
33          ur = delete(v_ur, 0, 1)         # delete v, 0 = leftmost
```

Here the second argument '0' in 'delete' stands for 'first', the third argument '1' again stands for 'column'.

This ends the iterative calculation in the for loop initiated in line 21.

When the computation is finished, process 1 returns its partial result 'ur' to the root process:

```
34  if rank == 1: send(ur, dest=0)
```

where it is appended by column_stack to the left-hand partial result computed by the root process itself:

```
35  if rank == 0:
36      ur = recv(source=1)
37      u = column_stack([ul, ur])
```

Finally, we let the root process plot the solution:

```
38  if rank == 0:
39      import matplotlib.pyplot as plt
40      from mpl_toolkits.mplot3d import Axes3D
41      from matplotlib import cm
42      fig = plt.figure()
43      ax = fig.gca(projection='3d')
44      from numpy import meshgrid
45      x, y = meshgrid(x,x)
46      surf = ax.plot_surface(x, y, u,
47                  rstride=1, cstride=1, cmap=cm.jet, linewidth=0)
48      plt.show()
```

The result is shown in Fig. 11.1 below.

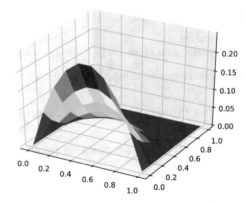

Fig. 11.1 Approximation to Laplace equation (1) in Sect. 11.5, with boundary function g in (2), computation distributed to 2 processes

11.6 Conjugate Gradient Method

In Sect. 4.2 in the Scientific Python chapter, we developed the conjugate gradient method for the solution of linear equation systems $Ax = b$ for symmetric and positive definite matrices A.

In the following, we show how the computation in the sequential program in Example 4.2 can be distributed to multiple processes.

Analysis of the sequential program shows that the by far largest computation effort is concerned with the matrix-vector products, hence in parallelization the main focus needs to be on the speed-up of matrix-vector multiplication.

The idea is to decompose A horizontally into contiguous blocks $A_i, i = 0, 1 \ldots, s - 1$, one for each of the s involved processses. A computation of Av for some vector v can then be partitioned into the partial computations $A_i v$, which are then concatenated to yield the total result. For simplicity we assume that s evenly divides the number n of rows of A.

The Program

As usual, the program begins with our mpi4py declarations. Then we import the NumPy components:

```
1 from numpy import array, zeros, sqrt
```

Next we introduce some variables which will be set in the root process. As mentioned, to access them also in the other processes, they must be declared globally. And, since in Python declaration without initialization is not possible, they receive the special value None:

```
2 A = n = p = None
```

Initial values are then set as in the sequential program, except that the commands are executed within the root process:

```
3 if rank == 0:
4     A = array([[  9.,    3.,   -6.,   12.], [  3.,   26.,   -7.,  -11.],
5               [ -6.,   -7.,    9.,    7.], [ 12.,  -11.,    7.,   65.]])
6     b = array([ 18.,   11.,    3.,   73.])
7     n = len(b)
8     x = zeros(n)
9     r = b.copy(); p = r.copy()
10    rs_old = r @ r
```

We partition A into horizontal slices, declare local matrices to store the individual submatrices and scatter A to the involved processes:

```
11 n = bcast(n)
12 n_loc = n//size
13 A_loc = zeros((n_loc, n))
14 Scatter(A, A_loc)
```

This distribution is required only once, before the actual computation begins.

The gradient descent is performed in the following for-loop in lines 15–27.
Each step begins with the distributed computation of the vector $A \cdot p$:

```
15 for i in range(n):       # n = max iterations to exact result
16     p = bcast(p)          # current p distributed to all delegates
17     Ap_loc = A_loc @ p    # local computation of matrix vector product
18     Ap = zeros(n)         # buffer needed to ...
19     Gather(Ap_loc, Ap)    # ... gather together local results in root
```

In line 16 the current value of the search direction vector p is broadcast to all processes. Each process computes its segment $A_{loc} \cdot p$ of the product $A \cdot p$ in line 17. In line 19 The segments are then collected in the root variable Ap.

The Gather operator is basically the converse of Scatter. It reassembles the local partial solutions to a total one in the root process, respecting the distribution order made by Scatter in line 14.

The root process can now set the new values for α, x, r and $r \cdot r$ in lines 21–24 below, and test whether the approximation is close enough. If yes, the iteration is terminated in line 25. Otherwise the next iteration round is prepared in lines 26 and 27.

Note that the if-block in lines 20–27 is indented, because it still belongs to the for loop:

```
20      if rank == 0:
21          alpha = rs_old / (p @ Ap)
22          x += alpha*p
23          r -= alpha*Ap
24          rs_new = r @ r
25          if sqrt(rs_new) < 1e-10: break
26          p = r + (rs_new / rs_old)*p
27          rs_old = rs_new   # end for Loop
```

After the loop is finished, the result is printed by root:

```
28  if rank == 0: print(x)    # Out: [1.0,   1.0,   1.0,   1.0]
```

To sum up, the only part that is executed in parallel is the computation of the next iteration values $A \cdot p$ in lines 16–19.

11.7 Exercises

Matrix Operations

Exercise 11.1 Write an mpi4py program to multiply two matrices A, B. Split the matrix B *vertically* into two slices B_1 and B_2, distribute the computation of AB_1 and AB_2 to *two* processes, and finally recombine the partial products and print the resulting product matrix.

Test the program for the Hilbert matrix of order 4 and its inverse. These matrices can be imported from scipy.linalg as hilbert(4) and invhilbert(4).

Exercise 11.2 Extend the program to one where additionally also A is partitioned into *two* parts A_1 and A_2, now however *horizontally*. Distribute the computation of the partial products $A_i B_j$ to *four* processes.

The alltoall Operator

The following exercise illustrates the use of the alltoall operator, which combines collection and distribution of data.

Assume that in a communication world, each of the 'size' many involved processes has a send buffer, say, sendbuf and a receive buffer, say, recvbuf with exactly 'size' data objects. The alltoall operator takes the *i*th object from the sendbuf of process *j* and copies it into the *j*th object of the recvbuf of process *i*.

The operation can be thought of as a transpose of the matrix with processes as columns and data objects as rows.

The syntax of the alltoall methods is

```
recvbuf = alltoall(sendbuf)
```

As with all 'lowercase' operators, the data objects can be of any allowed Python type, provided the entries in sendbuf conform to those in recvbuf.

Exercise 11.3 Write a program to compute the transpose of an 8×8 matrix A, say

```
A = arange(64).reshape(8,8)
```

distributed to 4 processes as follows:

Split the matrix vertically in 4 slices using the operator vsplit and scatter the slices to the distributed processes. Each process then subdivides the slice into a sequence of four 2×2 matrices using hsplit and transposes the submatrices locally with the NumPy operator transpose.

The resulting local matrix sequences are then processed with alltoall, and locally reassembled into an 8×2 matrix using the operator hstack.

The local column-slices are then gathered together in the root process, reshaped into an 8×8 matrix and finally printed.

The Sieve of Eratosthenes

Exercise 11.4 Write an mpi4py program sieve.py to find all prime numbers in the set $L = \{2, \ldots, n\}$.

The set L has to be divided into parts L_i and distributed to all processes *i*. Each process *i* searches for its local minimum p_i from its subset L_i and sends it to the root process.

The root process determines the global minimum $p = \min p_i$, and broadcasts it to all processes, which then delete all multiples of p from their subsets L_i. This step is to be repeated until the condition $p^2 \leq n$ becomes false.

Now each subset L_i only consists of prime numbers.

In the end, the largest prime number found should be printed to the console.

Test the program for $n = 100$, distributed to 4 processes.

Chapter 12
Parallel Computing in C/C++

MPI was originally developed for C, C++ and Fortran. In practice these programming platforms are still the most common. Based on the discussion in the last chapter, we briefly sketch how to implement message passing in C/C++. Actually, we
mainly concentrate on C, since on a basic level, message passing relies on elementary
C concepts such as pointers and arrays.

We then turn to another main paradigm in distributed computing, the *shared
memory* approach.

Installation

In the following we consider the widespread implementation *Open MPI* (not to be
confused with the shared memory module OpenMP, discussed later).

At the moment there does not seem to be a convenient approach for the development of Open MPI programs in standard integrated development environments
such as Eclipse or Xccode.

We base the discussion on the widespread operating system Ubuntu Linux, already mentioned in the C and C++ chapters. More precisely, we will compile and
run our programs in the Ubuntu Terminal application.

Ubuntu can either be installed directly or on a virtual machine within another
system. All examples in this chapter were tested under Ubuntu 18.04 in a virtual
environment in macOS.

In Ubuntu, Open MPI for C/C++ can be installed with the following Terminal command:

```
$ sudo apt-get install openmpi-bin openmpi-doc libopenmpi-dev
```

As a first example, assume `hello.c` to contain the usual introductory C program
from Example 6.1 in the C chapter, with the single `printf("Hello World\n")` command.

© Springer Nature Switzerland AG 2020
E. Smith, *Introduction to the Tools of Scientific Computing*, Texts in Computational
Science and Engineering 25, https://doi.org/10.1007/978-3-030-60808-8_12

The source code is compiled to an executable program `hello` with the MPI aware compiler `mpicc`:

```
$ mpicc hello.c -o hello
```

In Ubuntu, `mpicc` is just a a wrapper around the standard C compiler `gcc`, adding some components needed for parallel processing. Details can be inquired by

```
$ mpicc -showme
```

The corresponding C++ compiler can be called with `mpiCC`.

The option '`-o hello`' means that the compiled program is to be output as a file named 'hello'.

It can then be scheduled to, say, $n = 2$ processes and run with the command `mpiexec`:

```
$ mpiexec -n 2 hello
Hello World
Hello World
```

Note that in Python we had to supply '`python hello.py`' as an argument to `mpiexec`, which launched two instances of the *interpreter*. Here `hello` is an immediately executable standalone program.

The `mpi.h` Library

In Python we imported the central MPI module with the instruction '`from mpi4py import MPI`'. In C/C++, the command '`#include <mpi.h>`' provides the required communication components.

12.1 Integral Approximation of π

In Example 11.7 in the last chapter, we saw that π can be approximated by a midpoint computation of the integral

$$\int_0^1 \frac{4}{1 + x^2}\, dx.$$

Example 12.1 We write a corresponding parallel C program and store it in a file `parpi.c`.

In the preamble we first include the main MPI module and the standard input/output library:

```
1 #include <mpi.h>
2 #include <stdio.h>
```

We then define the function used in the integration:

```
3 float f(float x) { return 4.0/(1.0 + x*x); }
```

The function to compute the partial integral approximation:

```
4 float partial_pi(int n, int start, int step) {
5     float h = 1.0/n;
6     float sum = 0.0;
7     for (int i = start; i < n; i += step) {
8         float x = h*(i + 0.5);
9         sum += h*f(x); }
10    return sum; }
```

Note in the `main` function below that for the first time we need the arguments in parentheses, to pass the size of the communication world into the computation via the '-n' option:

```
11 int main(int argc, char *argv[]) {
```

The argument `argv` is picked up by the `MPI_Init`-command, which then provides the `size` and `rank` values as in Python:

```
12     MPI_Init(NULL,NULL);
13     int rank, size;
14     MPI_Comm_size(MPI_COMM_WORLD, &size);
15     MPI_Comm_rank(MPI_COMM_WORLD, &rank);
```

The root process then fixes the number n of interval segments and broadcasts it to all processes:

```
16     int n;
17     if (rank == 0) n = 10;
18     MPI_Bcast(&n, 1, MPI_INT, 0, MPI_COMM_WORLD);
```

Here we note a much more verbose specification. The arguments in line 18 are: the address of the buffer array to be broadcast, the number of entries (here 1, because only the n value is distributed), then the data-type `MPI_INT` of the buffer array, followed by the rank of the sender (here the root process 0), and finally the so-called *communicator*, i.e. the communication context.

Each process computes its partial contribution to the approximation:

```
19     float pi_loc = partial_pi(n, rank, size);
```

The local contributions are then added up by the root process:

```
20     float pi;
21     MPI_Reduce(&pi_loc, &pi, 1, MPI_FLOAT,
22                 MPI_SUM, 0, MPI_COMM_WORLD);
```

In lines 21 and 22 all local values `pi_loc` are sent to the root process and accumulated in `pi`. The arguments in position 3 and 4 declare that the send array consists of one element of type `MPI_FLOAT`. The next argument `MPI_SUM` specifies that the values are to be summed up in `pi`. The 0 value in the following argument signifies that the receiver is the process with rank 0. The last argument again denotes the communication context.

The value `pi` is then printed by the root process:

```
23      if (rank == 0) printf("%f\n", pi);
```

and the MPI computation terminated by the command

```
24      MPI_Finalize(); }
```

The finalize command is a required closing to the initialize command in line 12.

Remark This example indicates that MPI programming in C is not so different from Python, the main difference being that the commands are somewhat more verbose, since all arguments must be mentioned explicitly in C. Arguments are always identified by *position*, hence there is in particular no possibility to suppress the specification of default values.

12.2 Scatter and Gather

In the Python examples we repeatedly made use of the MPI commands `Scatter` and `Gather`. As a simple C example, we write a program to compute the average of a sequence of numbers.

Example 12.2 Assume the following program to be written to a file `avg.c`.
We begin as in the last example:

```
1  #include <mpi.h>
2  #include <stdio.h>
3  int main(int argc, char** argv) {
4      MPI_Init(NULL, NULL);
5      int size, rank;
6      MPI_Comm_size(MPI_COMM_WORLD, &size);
7      MPI_Comm_rank(MPI_COMM_WORLD, &rank);
```

We fix an array of length 16:

```
8      int n = 16;
9      float arr[n];
```

and let the root process fill it with numbers $0, 1, \ldots, 15$:

```
10     if (rank == 0) for (int i = 0; i < n; i++) arr[i] = i;
```

Then we prepare for the array to be divided into 'size' many parts, where we again for simplicity assume `size` to evenly divide n.

```
11     int n_loc = n/size;
12     float arr_loc[n_loc];
```

The input array is scattered evenly to all processes:

```
13     MPI_Scatter(arr, n_loc, MPI_FLOAT,
14                 arr_loc, n_loc, MPI_FLOAT, 0, MPI_COMM_WORLD);
```

The `MPI_Scatter` function may look scary, but it is actually only a more verbose formulation of the one in Python.

Next up is the computation of the averages of the numbers within each local process:

```
15    float sum_loc = 0.;
16    for (int i = 0; i < n_loc ; i++) sum_loc += arr_loc[i];
17    float avg_loc = sum_loc / n_loc;
```

The single values `avg_loc` are then gathered into the array `avg_arr` and sent to the root process 0:

```
18    float avg_arr[size];
19    MPI_Gather(&avg_loc, 1, MPI_FLOAT,
20                  avg_arr, 1, MPI_FLOAT, 0, MPI_COMM_WORLD);
```

The argument value '1' denotes that the transfer is performed via buffer arrays of length 1.

Finally, the root process computes the total average from the local averages, and then prints it:

```
21    if (rank == 0) {
22        float sum = 0.;
23        for (int i = 0; i < size ; i++)
24            sum += avg_arr[i];
25        float avg = sum/size;
26        printf("%f\n", avg); }
```

The program ends with the closing statement:

```
27    MPI_Finalize(); }
```

12.3 Conjugate Gradient

As a larger "real world example", we develop a C implementation of the parallel conjugate gradient method introduced in Sect. 4.2 in the SciPy chapter, then parallelized in Python in Chapt. 11.

Recall that there are two procedures in the algorithm that could gain from parallelizing: vector-vector multiplication and matrix-vector multiplication. As in the Python version, we only consider the distributed computing of the latter.

Example 12.3 Assume the following program is stored in a file `parconjgrad.c`. We start with the preamble:

```
1 #include <stdio.h>
2 #include <mpi.h>
3 #include <math.h>   // contains sqrt, see compilation remark below
```

The following function (sequentially) computes the dot product between two vectors:

```
4 float dot(int n, float v[], float w[]) {
5     float sum = 0;
6     for (int i = 0; i < n; i++) sum += v[i]*w[i];
7     return sum; }
```

In the main function, we first fix our example values A and b and the start value 0 for the solution vector x:

```
 8 int main(int argc, char *argv[]) {
 9     int n = 4;
10     float A[] = {9., 3., -6., 12., 3., 26., -7., -11.,
11                  -6., -7., 9., 7., 12., -11., 7., 65.};
12     float b[] = {18., 11., 3., 73.};
13     float x[] = {0., 0., 0., 0.};
```

Note that in lines 10 and 11 we initialize A in a flattened form as a linear vector. There are two reasons. First, C does not permit to use the variable n from line 9 to declare a matrix in the form 'A[][n]='. Second, we want to scatter A in contiguous blocks of rows. For that we need the flattened form anyway, since transfer buffers can only hold arrays.

Then we initialize the residual vector r, search-direction p and dot product $r \cdot r$:

```
14     float r[n], p[n];
15     for (int i = 0; i < n; i++) { r[i] = b[i]; p[i] = r[i]; }
16     float rs_old = dot(n,r,r);
```

The communication components:

```
17     int rank, size;
18     MPI_Init(&argc, &argv);
19     MPI_Comm_size(MPI_COMM_WORLD,&size);
20     MPI_Comm_rank(MPI_COMM_WORLD,&rank);
```

We partition A into horizontal slices, declare local matrices to store the individual submatrices and scatter A to the involved processes:

```
21     int n_loc = n/size;
22     float A_loc[n_loc][n];
23     MPI_Scatter(A, n_loc * n, MPI_FLOAT,
24                 A_loc, n_loc*n, MPI_FLOAT, 0, MPI_COMM_WORLD);
```

This distribution is required only once, before the actual computation begins.

The iterative approximation is performed in the following for loop from line 25 to 44:

```
25     for (int it = 0; it < n; it++) {    // begin iteration loop
```

Each step begins with the broadcast of the current value of the search direction vector p to all processes:

```
26         MPI_Bcast(p, n, MPI_FLOAT, 0, MPI_COMM_WORLD);
```

Each process computes its segment $A_{loc} \cdot p$ of the vector $A \cdot p$:

```
27         float Ap_loc[n/size];
28         for (int i = 0; i < n_loc; i++) {
29             float sum = 0;
30             for (int j = 0; j < n; j++) {
31                 sum += A_loc[i][j]*p[j];}
32             Ap_loc[i] = sum;}
```

The segments are then gathered together in an array `Ap` in the root process:

```
33          float Ap[n];
34          MPI_Gather(Ap_loc, n_loc, MPI_FLOAT,
35                     Ap, n_loc, MPI_FLOAT, 0, MPI_COMM_WORLD);
```

The root process can now set the new values for α, x, r and $r \cdot r$:

```
36          if (rank == 0) {
37              float alpha = rs_old / dot(n, p, Ap);
38              for (int i = 0; i < n; i++)  x[i] += alpha*p[i];
39              for (int i = 0; i < n; i++)  r[i] -= alpha*Ap[i];
```

If the approximation is close enough, the iteration is terminated:

```
40              float rs_new = dot(n, r, r);
41              if (sqrt(rs_new) < 1e-6) break;
```

Otherwise the next iteration round is prepared:

```
42              for (int i = 0; i < n; i++)
43                  p[i] = r[i] + (rs_new / rs_old)*p[i];
44              rs_old = rs_new; } }    // end if, end iteration loop
```

After the loop is finished, the result is printed by root and the MPI session is ended:

```
45      if (rank == 0)  {
46          for (int i = 0; i < n; i++)  printf("%f ", x[i]);
47          printf("\n"); }
48      MPI_Finalize(); }
```

Compilation and Testing

As mentioned before, the compiler in Ubuntu is somewhat peculiar with respect to the library `math.h`. We have to explicitly include the '-lm' option after the compile command:

```
$ mpicc parconjgrad.c -o parconjgrad -lm
$ mpiexec -n 2 parconjgrad
0.999984 1.000027 0.999988 0.999893
```

12.4 Shared Memory Programming

If the message passing model in distributed computing corresponds to independent co-workers communicating by, say, email, by analogy, in a shared memory model, multiple workers crowded around a common table all operate on the same data.

In more technical terms, the shared memory approach is mainly targeted at parallelism *within* a (multi-core) unit, while message passing is used for communication between *separate computing nodes*.

For shared memory programming, *OpenMP* is the de facto standard. Here "*MP*" stands for "*Multi Processing*" and *not* for *Message Passing*, as in MPI.

OpenMP supports shared-memory multiprocessing programming in C, C++ and Fortran, however not in standard Python, where a so-called *global interpreter lock* prevents concurrent code execution.

The development of OpenMP began in the late 1990s, and has since then been implemented in many compilers, commercial as well as open source. In the following we base the discussion on C, specifically on the implementation in Ubuntu Linux, where all the needed components are already included.

Central to OpenMP is the idea of *threads* that exist as components of a process and can be executed independently. More specifically, OpenMP is an implementation of *multithreading*, a parallelization method in which a master thread branches off a certain number of worker threads and the system divides a task among them. The threads then run concurrently, with the runtime environment allocating threads to different processors.

Hello World

We illustrate the main ideas with a simple "Hello World" program in C:

Example 12.4 We assume the following code to be stored in a file hellomp.c:

```
1 #include <stdio.h>
2 #include <omp.h>
3 int main() {
4     printf("Number of processors: %d \n", omp_get_num_procs());
5     #pragma omp parallel
6     { printf("Hello World from thread number %d of %d\n",
7              omp_get_thread_num(), omp_get_num_threads()); } }
```

Line 2 provides the OpenMP environment-variables and -functions used in the program.

In line 4 the function omp_get_num_procs returns the number of processors available in the system.

The #pragma directive in line 5 is the method specified by the C standard for providing additional information to the compiler, beyond what is conveyed in the language itself. Here the pragma 'omp parallel' tells the compiler that the commands in the following block consisting of lines 6 and 7 should be executed in parallel by all threads, which by default means one thread per processor.

The default value can be changed by num_threads(n) for the desired number *n* of threads.

Compilation and Program Execution

To compile the program under Ubuntu Linux, assuming the file to be stored in the home directory, we write the command line

```
$ gcc -fopen hellomp.c -o hellomp
```

The option '-fopen' enables the handling of OpenMP directives '#pragma omp'. When it is specified, the compiler generates parallel code according to the OpenMP Application Program Interface (API).

If the MPI interface is already installed, we can write `mpicc` instead of `gcc`, with the same effect.

We run the program on a machine with 8 processor units:

```
$ hello
Number of processors: 8
Hello World from thread number 7 of 8
...
Hello World from thread number 2 of 8
```

The ellipsis represents the six remaining responses from the other threads.

Message Passing with Shared Memory

Note the similarity between threads in OpenMP and processes in MPI. In fact, we can also program in "message-passing style" in OpenMP. As a simple example, we show how to implement a broadcast operation.

Example 12.5 The following program is intended to broadcast the value 42 from thread number 0 to all threads:

```
1 #include <stdio.h>
2 #include <omp.h>
3 int main() {
4     int ans = 0;
5     #pragma omp parallel
6     { if (omp_get_thread_num() == 0) ans = 42;
7       printf("%d ", ans); }
8     printf("\n"); }
```

The code block in lines 6 and 7 is executed by all threads concurrently. The only effect of the statement in line 6 is that thread 0 assigns the value 42 to the shared variable `ans`. In line 7 the current value of `ans`, as seen by the individual threads, is printed. Line 8 is outside of the `parallel` block, so that the line-break is performed only at the end of the program.

The result is however not entirely satisfactory. It may randomly change between program executions, a typical output being

```
0 0 42 0 0 42 42 0
```

The reason is again that the block is executed *concurrently*. It is perfectly possible that, say, thread 1 executes line 7 *before* thread 0 changes the value of `ans`. In that case, thread 1 prints the value 0. If however thread 1 executes line 7 *after* the value change, then the expected value 42 is printed.

The barrier Directive

To guarantee the desired broadcast behavior, we have to enforce synchronization between thread executions, such that each thread waits with the execution of line 7 until *after* thread 0 has stored the value 42 into ans. This can be achieved with the OpenMP 'barrier' directive.

Example 12.6 If we insert the line

```
#pragma omp barrier
```

between lines 6 and 7 in the last example, then each thread suspends its execution until also all the others have processed line 6, i.e. in particular until thread 0 has updated ans, and only then resumes the execution of line 7. Then the result will be the desired one:

```
42 42 42 42 42 42 42 42
```

12.5 Parallelizing Loops

A good candidate for parallel execution is a suitably structured for loop. For instance, it is obvious that in a vector addition like

```
for (int i = 1; i <= n; i++) c[i] = a[i] + b[i];
```

the component additions are independent of each other, and can thus be executed in parallel.

For the distribution of a for loop to separate threads, OpenMP provides the directive

```
#pragma omp for
```

Combined with the `parallel` directive, the program snippet to initiate a parallel vector addition looks like this:

```
1 #pragma omp parallel
2    {
3        #pragma omp for
4        for (int i = 1; i <= 100; i++) c[i] = a[i] + b[i];
5    }
```

Recall that the code block to be computed in parallel is as usual enclosed in curly braces (lines 2 and 5). Again if the block consists of a single statement, the braces can be left out.

The combination of the two directives above occurs so often, that a short form is provided:

```
#pragma omp parallel for
for (int i = 1; i <= 100; i++) c[i] = a[i] + b[i];
```

It is important to note that the compiler does not check whether it is safe to parallelize the loop. Incautious use can lead to rather subtle errors, as we can see in the following example:

Example 12.7 Assume we want to sum up the first 100 natural numbers. In a sequential program we could do it with the following code snippet inside the `main` function:

```
int sum = 0;
for (int i = 1; i <= 100; i++) sum += i;
```

In a first attempt to parallelize it, we proceed as above:

```
1 #include <stdio.h>
2 int main() {
3     int sum = 0;
4     #pragma omp parallel
5     {
6         #pragma omp for
7         for (int i = 1; i <= 100; i++) sum += i;
8     }
9     printf("%d\n", sum); }
```

If we save the program in a file `formp.c`, compile it with the option '`-fopenmp`' and run it, then the result printed will more often than not be the expected:

```
$ formp
5050
```

However, the correct result is not guaranteed. It may happen that for instance

```
$ formp
4983
```

How is such a result to be explained? It is due to the most intricate and subtle error source in shared memory computation, the so called *race condition*.

The Bane of Shared Memory Computation: Race Conditions

In Example 12.5 above we saw that the order of execution between statements in independent threads can produce undesirable random results. Actually, on a more fundamental micro level, this is precisely what also happens in Example 12.7.

To see this, we consider a simplified version. Again let `sum` = 0 and let thread 0 increment `sum` by 1, thread 1 increment it by 2.

Recall that an assignment is a *sequence of three parts*, (1) accessing of the old value, (2) processing that value, and finally (3) storing the new value.

Since the threads are executed concurrently, it cannot be excluded that the sequences are interleaved, such that first both threads read the value 0, then increment the value to 1 or 2, and finally write it back to `sum`. Depending on which of the threads is the last to write its computed value to `sum`, the result will be either 1 or 2, but in both cases not the correct value 3.

The final result of this code depends on which one of the threads finishes last, and therefore writes to memory last, which is a race condition.

The critical Directive

There are several ways to get around a race condition.

The 'omp critical' directive identifies a section of code that must be executed by a single thread at a time. In our Example 12.7 we could replace line 7 by

```
10 for (int i = 1; i <= 100; i++) {
11     #pragma omp critical
12         sum += i;
13 }
```

to exclude interleaved execution of the assignments.

However in the present case we would loose all advantages of parallelizing, returning in fact to sequential execution of the loop. The critical directive is often used to ensure exclusive access to single resources such as printers or the output to the screen.

The atomic Directive

The atomic directive can be seen as a particular form of the critical directive, which specifically only prevents an *assignment* from being interrupted at any stage by another thread. So here

```
14     #pragma omp atomic
```

instead of line 11 would work perfectly.

The performance of atomic is generally significantly better than that of critical. However, as with the critical directive, we incur the cost of serialization.

In our particular case it is still not the best option.

The reduction Clause

The best method to use is 'reduction'. We can do this by changing line 6 in Example 12.7 to

```
15     #pragma omp for reduction(+:sum)
```

The reduction command tells OpenMP that we want each thread to keep track of its own private sum variable while the loop is running, and add them all up at the end of the loop. This is the most efficient method, as the entire loop now runs in parallel, with the only overhead being right at the end of the loop, when the local sum values are added.

Similarly to the reduce method in MPI, the reduction directive can also be used with various other arithmetical operators, such as *, max, min.

12.6 Integral Approximation of π

As a first real world example we return to the integral approximation of π with the midpoint method in Example 12.1.

Example 12.8 Here is the complete program, first the preamble, which includes the definition of the function $f(x) := 4/(1 + x^2)$ to be integrated:

```
1 #include <stdio.h>
2 float f(float x) { return 4.0/(1.0 + x*x); }
```

The main block defines the number n of equidistant grid points and the distance h between them:

```
3 int main() {
4     int n = 10; float h = 1.0/n;
```

The variable x is used to range over the discrete evaluation points, the piecewise integral approximations are added up in sum:

```
5     float x;   // every thread gets its own copy, see line 9
6     float sum = 0.0;
```

The rest of the program is distributed to all threads:

```
7     # pragma omp parallel
8     {              // parallel block up to line 14
```

In the following loop, the partial integrals are computed concurrently and added up in sum. The variable x is declared as private, which means that each thread receives a separate memory space to store its value.

```
9         #pragma omp for private(x) reduction(+:sum)
10            for (int i = 0; i < n; i++) {
11                x = h*(i + 0.5);
12                sum += h*f(x); }
```

It remains to print the result:

```
13        #pragma omp single
14            printf("%f\n", sum); } }
```

In MPI we usually let the root process (i.e. the one with rank 0) print the result. Here we leave it to an arbitrary thread. The thread that reaches the single directive in line 13 first, is the one that executes the `single` block, i.e. the print instruction in line 14.

Note that the for-directive in line 9 automatically ensures that control is only passed to the commands that follow the loop when all threads have ended. The effect is the same as if there were an explicit `barrier` directive after line 12.

12.7 Parallelized Function Modules

So far the parallelization was organized in the `main` function. However it is also possible to outsource the parallel code to external function modules.

As a first example we define a function omp_dot that computes the dot vector product in parallel. It can then be used within an otherwise sequential program.

Example 12.9 After the usual

```
1 #include <stdio.h>
```

we define a function for the distributed computation of the vector dot product:

```
2 float omp_dot(int n, float v[], float w[]) {
3     float sum = 0;
4     #pragma omp parallel for reduction(+:sum)
5         for (int i = 0; i < n; i++) sum += v[i]*w[i];
6     return sum; }
```

The parallel dot product is then tested inside a sequential main module:

```
7  int main() {
8      int n = 10;
9      float a[n], b[n];
10     for (int i=0; i < n; i++){
11         a[i] = i; b[i] = n-i; }
12     float a_dot_b = omp_dot(n, a, b);   // computed in parallel
13     printf("%f\n", a_dot_b); }
```

The program must of course be compiled with the '-fopenmp' option for the parallelizing to take effect:

```
$ gcc -fopenmp dotmp.c -o dotmp
$ dotmp
165.000000
```

Recursive Functions

Recursive functions can also be formulated in parallel code and outsourced to external modules. Here the OpenMP directive 'task' can be used to schedule recursive calls in a "to do list".

Example 12.10 As a typical example we consider the Fibonacci function.
We begin with the function module:

```
1  int fib(int n) {
2      int i, j;
3      if (n < 2) return 1;
4      #pragma omp task shared(i)
5          i = fib(n-1);
6      #pragma omp task shared(j)
7          j = fib(n-2);
8      #pragma omp taskwait
9          return i + j; }
```

The call to fib(n) generates two tasks, indicated by the task directive. One of the tasks computes fib(n-1) and the other computes fib(n-2). The return values are added together to produce the value returned by fib(n).

Each of the calls to fib(n-1) and fib(n-2) will in turn generate two tasks. Tasks will be recursively generated until the argument passed to fib() is less than 2.

The tasks will then be executed in reverse order, storing the intermediate results in the variables i and j.

The taskwait directive ensures that the two tasks generated in an invocation of fib are completed (that is, i and j have received their values) before that invocation of fib returns.

The function is then called like this:

```
10 #include <stdio.h>
11 int main() {
12     int n = 20;
13     #pragma omp parallel
14     {
15         #pragma omp single
16             printf("%d\n",fib(n)); } }
```

The parallel directive in line 13 specifies that the following code block is to be executed by all threads. The single directive in line 15 states that actually only one of the threads *initiates* the computation and prints the result. However, the computation itself is spread across all threads.

12.8 Hybrid Message Passing and Shared Memory Programming

It is perfectly possible to write hybrid parallel programs, that take advantage of both message passing and shared memory.

As a first example we reconsider the π approximation in Examples 12.1 and 12.8.

Assume the program is stored in a file parmp.c.

We begin as before:

```
1 #include <stdio.h>
2 #include <mpi.h>
3 float f(float x) { return 4.0/(1.0 + x*x); }
```

In Example 12.1 we used a (sequential) function partial_pi to compute the partial integral approximations. Here we distribute the for loop used in that function to all threads, and arrive at the function:

```
4 float omp_partial_pi(int n, int start, int step) {  // shared memory
5     float h = 1.0/n;
6     float sum = 0.0;
7     #pragma omp parallel for reduction(+:sum)
8         for (int i = start; i < n; i += step)
9             sum += h*f(h*(i + 0.5));
10     return sum; }
```

In the following main function, the only change from Example 12.1 is that our new parallel helper function omp_partial_pi replaces the sequential partial_pi. For convenience we repeat the complete definition:

```
11  int main(int argc, char *argv[]) {
12      MPI_Init(NULL,NULL);
13      int rank, size;
14      MPI_Comm_size(MPI_COMM_WORLD,&size);
15      MPI_Comm_rank(MPI_COMM_WORLD,&rank);
16      int n;
17      if (rank == 0) n = 10;
18      MPI_Bcast(&n, 1, MPI_INT, 0, MPI_COMM_WORLD);
19      float pi_loc = omp_partial_pi(n, rank, size);  // shared memory
20      float pi;
21      MPI_Reduce(&pi_loc, &pi, 1, MPI_FLOAT,
22                 MPI_SUM, 0, MPI_COMM_WORLD);
23      if (rank == 0) { printf("%f\n", pi); }
24      MPI_Finalize(); }
```

Since the program contains MPI as well as OpenMP components, it must be compiled with the extended gcc compiler mpicc using the '-fopenmp' option, and then be executed by mpiexec:

```
$ mpicc -fopenmp parmp.c -o parmp
$ mpiexec -n 2 parmp
3.142426
```

Hybrid Conjugate Gradient

As a final example, we provide a hybrid message-passing and memory-sharing parallel program for the computation of the conjugate gradient. The idea is to (1) distribute matrix-vector multiplication to different processors according to the message passing interface, and (2) distribute vector-vector multiplication to multiple threads.

In this case, all the work is already done. In the program code in Example 12.3, we only have to replace the helper function 'dot' in lines 4–7 with the shared value version 'omp_dot' in lines 2–6 in Example 12.9.

Assume the program is stored in a file conjgradmp.c. It can then be compiled and run as follows:

```
$ mpicc -fopenmp conjgradmp.c -o conjgradmp -lm
$ mpiexec -n 2 conjgradmp
0.999984 1.000027 0.999988 0.999893
```

12.9 Exercises

Message passing in C/C++

The following exercises refer to the exercise section in Chapt. 11.

Exercise 12.1 Write a C-version of the program in Exercise 11.1.

Exercise 12.2 Write a C-version of the program in Exercise 11.2.

Exercise 12.3 Write a C-version of the program in Exercise 11.3.

Exercise 12.4 Write a C++ version of the program in Exercise 11.4. Here you might consider to use the C++ data type vector to store the elements.

Shared Memory

Exercise 12.5 Based on the program in Example 6.8 in the C chapter, write a recursive function det(n,A) that computes the determinant of an $n \times n$ matrix A distributed to 8 threads.

Hybrid Message Passing - Shared Memory Programming

Exercise 12.6 Write a distributed program to compute the maximum norm $||A||_\infty = \max_{i,j} |a_{i,j}|$ of a matrix $A = (a_{i,j})$. Partition A into submatrices $A_1, \ldots, A_m$, and distribute the parts to separate MPI-processes. In each process, distribute the computation of the local maximum to multiple threads.

Chapter 13
Distributed Processing in Julia

Not surprisingly, as a modern language for high-performance computing, Julia offers on-board support for parallel and distributed computing. While it is possible to integrate external environments such as MPI, Julia's proprietary module `Distributed` in the standard library proves to be a versatile alternative. In this chapter we give a brief introduction to this latter approach.

Basics

If Julia is already launched, the `Distributed` module is loaded by

```
using Distributed
```

We then set the number of desired additional processes:

```
addprocs(6)
```

For best performance, the documentation recommends to choose the number n of physically available processors.

If access to the Julia application has been enabled according to Remark 8.1 in the Julia chapter, Julia can be launched from the Terminal like this:

```
$ julia -p 6
```

which then automatically loads the `Distributed` module and initializes it with `addprocs(6)`.

In any case we get the following:

```
nprocs()    # number of available processes: 7
procs()     # their identification numbers [1, 2, 3, 4, 5, 6, 7]
nworkers()  # number of work processes: 6
workers()   # [2, 3, 4, 5, 6, 7]
```

In `Distributed`, process 1 is assumed to be the *main* process, which controls communication with the remaining so-called *worker* processes.

© Springer Nature Switzerland AG 2020
E. Smith, *Introduction to the Tools of Scientific Computing*, Texts in Computational Science and Engineering 25, https://doi.org/10.1007/978-3-030-60808-8_13

13.1 Point-to-Point Communication

In Julia the distinguished main process 1 plays a similar role as the root process in MPI.

Basic communication is initiated by the function `remotecall` that delegates a function (in the example below the square-root function) together with an argument (here 2.0) to a worker (here worker 2), and retrieves the result like this:

```
r = remotecall(sqrt, 2, 2.0);
fetch(r)  # ans: 1.4142135623730951
```

The `fetch` command extracts the return value for `sqrt(2)`.

The pair `remotecall`/ `fetch` corresponds to `send`/`receive` in MPI.

In general, it is more convenient to use the macro `@spawnat`, which works in the same way except that it takes an *expression* instead of a function:

```
r = @spawnat 2 sqrt(2.0)
```

Also a block of code can be delegated by enclosing it in a `begin`/`end` pair:

Example 13.1 We consider the task of computing $\sqrt{2}$ by Newton approximation delegated to worker 3:

```
1 sq = @spawnat 3 begin
2       r = 1; eps = 1.e-6
3       while abs(r^2 - 2) > eps   r = (r + 2/r)/2 end
4       return r
5 end;
6 fetch(sq)   # ans: 1.4142135623746899
```

There is also a variant '`@spawn`' of '`@spawnat`' that leaves the choice of worker to the system.

Example:

```
r = @spawn sqrt(2.0)
```

13.2 Distributed Loops, Reductions

So far we have considered direct point-to-point communication. Very convenient are however also techniques to distribute tasks over multiple processes. Of central importance are iteration loops, where the partial results are combined using a specified reduction method. This corresponds to the reduction concept in OpenMP, discussed in Sect. 12.4 in the last chapter.

Example 13.2 Here is the computation of the approximation $\sum_{i=1}^{10} 1/2^i \approx 1$:

```
@distributed (+) for i = 1:10   1/2^i end   # ans: 0.9990234375
```

The macro `@distributed` declares that the term evaluations in the loop should be distributed to all workers. The '(+)' reduction parameter specifies that the individual results should be combined and their total sum returned.

Similarly, the factorial 10! can be computed by

```
@distributed (*) for i = 1:10  i end  # ans: 3628800
```

Remark We mention that in the examples above, index values are distributed to the worker processes in a way that automatically takes care of load balancing, following the same pattern as described for MPI in the previous chapters.

This can be seen e.g. from the loop

```
@distributed for i = 1:10 println("index $i processed here") end;
```

where the `println` command outputs 10 lines of the form

```
From worker 4:  index 5 processed here
```

Example 13.3 A "real-world" example for the use of distributed for-loops is the computation of the vector dot-product $v \cdot w$:

```
v = collect(1. : 10.);  w = collect(10.: -1.: 1.);  # example vectors
@distributed (+) for i = 1:10 v[i]*w[i] end          # ans: 220.0
```

13.3 Monte Carlo Method

Monte Carlo methods are generally good candidates for distributed implementation. We illustrate the idea with our standard example, the approximation of π.

Example 13.4 The first approach is to formulate a sequential program, which is then distributed to all workers, and the partial results then recombined.

Here is the sequential program from Example 8.4 in the Julia chapter, embedded in a function:

```
1 function pi_loc(samples)
2     hits = 0
3     for _ = 1:samples
4         x, y = rand(), rand()
5         if(x^2 + y^2 <= 1) hits += 1 end end
6     return 4*hits/samples end
```

For later comparison with the distributed computation we time the performance:

```
7 samples = 100_000_000;
8 @elapsed pi_loc(samples)  # ans: 1.155978
```

We want to distribute the computation to all worker processes. Here is a naive approach:

```
9  function pi_dist(samples, p)
10     r = @distributed (+) for i = 1:p  pi_loc(samples/p) end
11     return r/p end
```

We test it:

```
12 p = nworkers()
13 pi_dist(samples, p)  # ans: Error!
```

The detailed error message states that `pi_loc` is unknown in the worker processes.

The reason is that the function definition in its current form is only accessible in the main process 1 and not, as required in line 10, also by the workers.

To make it generally applicable, we need to include line 1 in the macro @everywhere. The new line 1 now reads:

```
@everywhere function pi_loc(samples)  # now distributable
```

The macro @everywhere ensures that the enclosed code block is distributed also to all worker processes.

Now the command `pi_dist(samples,p)` in line 13 works.

A timing shows that there is a considerable performance gain, but how large it is, will of course depend on the number of physical processors. On our test machine with 8 cores we get:

```
14 @elapsed pi_dist(samples, p)  # ans: 0.164849371
```

Compared to the sequential program in line 8, we get a speed increase by a factor of about 7.

Example 13.5 In the example above we distributed the given sequential program. In a simple case like here, where there is no inter-process communication between the partial computations, we can actually specify the distributed computation directly in a single for loop:

```
1 samples = 100_000_000;
2 hits = @distributed (+) for _ = 1:samples
3             Int(rand()^2 + rand()^2 <= 1) end;
4 mypi = 4*hits/samples
```

Note that line 3 simply combines the effect of lines 4 and 5 in the sequential program. The expression 'rand()^2+rand()^2 <= 1' results in true or false, which is then converted into an integer 1 or 0 that can be summed up. (Actually, the explicit conversion is even redundant, since Boolean values correspond to integers 1 and 0 anyway.)

If you want to time the execution, you can enclose lines 2 and 3 in a block:

```
@elapsed begin
 # code goes here
end
```

13.4 Fibonacci Function

In Example 12.10 in the last chapter, we discussed an OpenMP program for the recursive computation of the Fibonacci function. Here is a corresponding distributed Julia implementation:

Example 13.6 We begin with the standard sequential program, controlled by the @everywhere macro, such that the function fib is accessible also to all workers.

```
1  @everywhere function fib(n)
2      if (n < 2) return n end
3      return fib(n-1) + fib(n-2)
4  end
```

Based on the sequential computation we formulate a distributed version as follows:

```
5  @everywhere function fib_parallel(n)
6      if (n < 40) return fib(n) end
7      x = @spawn fib_parallel(n-1)
8      y = fib_parallel(n-2)
9      return fetch(x) + y
10 end
```

Note that here the @everywhere macro is required to enable recursive function calls by the worker processes in line 7.

In distributed computation there is generally a trade off between performance gain due to multiple processors, and a slowdown caused by communication overhead, when data has to be moved around.

In the Fibonacci example, a compromise is to compute values for small n within a single process, as in line 6. Moreover, note that only one of the branches of the recursion calls is spawned to another process.

For the input value $n = 50$, we get a performance gain of 6.61 on our 8-core test machine:

```
11 @elapsed fib(50)             # ans: 53.926451246
12 @elapsed fib_parallel(50)    # ans:  8.156753204
```

13.5 Shared Arrays

As seen in Example 13.3, Julia arrays that are defined in the main process can also be read by the distributed workers. Workers can however not *modify* standard arrays of type Array. Now, Julia offers the special data type SharedArray for arrays that can be accessed by the workers also in writing mode.

The SharedArray type is provided in the module SharedArrays. In the following we assume it to be loaded by

```
using SharedArrays  # loads the required module
```

The first example illustrates the difference between the array types:

Example 13.7 We first define a standard Julia array

```
1 v = [0, 0];
```

We convert it to an array of type SharedArray with Julia's generic 'convert' function:

```
2 sv = convert(SharedArray, v);
```

We modify the values of both arrays in a worker process:

```
3 @spawn sv[1] = v[1] = 1;
```

In case of the standard array v, the new value is only assigned to a local copy, not visible from outside. For the shared array sv however, the change is global:

```
4 println("$v, $sv")   # ans: [0, 0], [1, 0]
```

Example 13.8 (Laplace Equation) Shared arrays provide a convenient way to implement shared memory programming. To illustrate, we write a distributed program to compute the Laplace equation, formulated in Python in Sect. 11.5.

Assuming the module SharedArrays is loaded, we define a shared matrix S to store the iterative approximations:

```
1 n = 10
2 S = SharedArray{Float64,2}((n,n));
```

Here the argument '2' denotes that S is a 2-dimensional array, i.e. a matrix, the tuple (n,n) that it is a $n \times n$ matrix. Note that S is automatically 0-initialized.

Still on the global level, we work the boundary conditions into the matrix:

```
3 x = collect(range(0, stop=1, length=n));
4 S[1,:] = x .* (1 .- x);
```

The iterative computation is embedded in a for loop as before, here however a distributed loop:

```
5  for _ = 1 : 40
6      T = copy(S)
7      @sync @distributed  for i in 2:n-1
8          for j in 2:n-1
9              S[i,j] = (T[i+1,j] + T[i-1,j] + T[i,j+1] + T[i,j-1]) / 4
10          end
11      end  # of distributed loop
12 end
```

In line 6 the values in S from the last iteration round are stored in the temporary matrix T. Note that we *copy* the values to avoid side effects.

In line 7 we initiate the for loop to update each grid point. As usual, the @distributed macro ensures that the computation is spawned to all workers. The @sync macro has the effect that the next iteration round is only started after all worker contributions to the current round have been included in S. (Note that in the *reduction* loops above, this synchronization happens automatically.)

The solution can then be plotted:

```
13 using Plots
14 plot(x, x, S, seriestype=:wireframe)
```

The result is illustrated in Fig. 13.1.

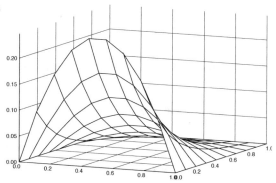

Fig. 13.1 Solution to Laplace equation in Sect. 11.5 in Chapt. 11. Computation distributed to 6 worker processes

13.6 Distributed Arrays

Distributed arrays are in a sense the converse of shared arrays. The idea is to split a given array and send the parts to different workers for further processing.

We need the module `DistributedArrays`. It is not included in the Julia standard library, so we must install it before first use:

```
import Pkg;                       # Julia's package manager
Pkg.add("DistributedArrays")  # only required once
```

From then on it can be used like any built-in module.

We present a simple example to see how distributed arrays work:

Example 13.9 We want to have access to the distributed arrays also in the worker processes, so we load the module in the form

```
1 @everywhere using DistributedArrays
```

We start with a standard Julia array

```
2 u = collect(1:10);
```

and convert it to a distributed array du:

```
3 du = distribute(u);
```

The array du is prepared to be split into local parts, one for each worker process:

```
4 for i in workers()
5     r = @spawnat i localpart(du)
6     lu = fetch(r)
7     println("$i: $lu")
8 end;
```

In line 5 each worker extracts its own part and prepares it for delivery in the variable r. Note that the localpart function belongs to the module DistributedArrays, and can be accessed by the workers only because the module has been loaded under the control of the @everywhere macro.

In line 6 the local part is fetched by the main process, and printed in line 7. The output (here collected in one line)

```
2: [1, 2]   3: [3, 4]   4: [5, 6]   5: [7, 8]   6: [9]   7: [10]
```

shows that the standard pattern of load balance is once again followed automatically.

Remark If the number of workers exceeds the length of the vector u, the last workers receive an empty part, such as for instance for u = collect(1:4):

```
2: [1]   3: [2]   4: [3]   5: [4]   6: Int64[]   7: Int64[]
```

Vector Dot Product

Example 13.10 As a mathematically more interesting example, we define a function for the distributed computation of the scalar dot product of two vectors:

```
1 function dot(u,v)
2     s = 0.0
3     du = distribute(u); dv = distribute(v)
4         for i in workers()
5             r = @spawnat i localpart(du)'*localpart(dv)   # note transpose
6             s += fetch(r)
7         end
8     return s
9 end
```

Recall that vectors are conceived as column vectors. For the dot multiplication in line 5 to go through, the left vector must be transposed to a row vector.
    We test the function:

```
u = collect(1. : 10.); v = collect(10.: -1: 1.0);
dot(u,v)   # ans: 220.0
```

Matrix-Vector Product

Example 13.11 We extend the example to a function to compute matrix-vector products Av. The idea is to split the matrix A horizontally into slices A_i, then to compute the $A_i v$ locally, and finally concatenate the partial-result vectors.

Here we have to take care if the number n of rows in A, in Julia written as size(A,1), is less than the number of workers. In that case we distribute the computation only to the first n workers, i.e. the processes $2, \ldots, n + 1$.

```
 1  function mat_vec(A,v)
 2      w = Float64[]                        # empty vector
 3      m = min(size(A,1), nworkers())       # see remark above
 4      dA = distribute(A, dist=[m, 1])
 5      for i = 2: m+1
 6          r = @spawnat i localpart(dA)*v
 7          w = vcat(w, fetch(r))           # append to w
 8      end
 9      return w
10  end
```

In line 2 we start with an initially empty vector, to which the partial product vectors are then appended. In line 3 we fix the number of worker processes, according to the remark above.

In line 4 the option dist specifies the number of partitions in each dimension. We split the matrix into m horizontal slices. The second entry '1' means that no splitting of the matrix along the column dimension is requested.

In line 6 the product of the local matrix slice and the vector v is computed in each of the involved workers and the partial result is then appended to the vector w in line 7.

Example 13.12 (Conjugate Gradient) As an application example for the distributed functions dot and mat_vec defined above, we write a program to solve matrix equations $Ax = b$ with the *conjugate gradient method* discussed before.

For introduction to the method see Sect. 4.2 in the SciPy chapter.

Here we formulate a distributed variant of the sequential Julia program from Example 8.18 in the Julia chapter.

We first fix the matrix A and the vector b as before:

```
 1  A = [9. 3. -6. 12.; 3. 26. -7. -11.; -6. -7. 9. 7.; 12. -11. 7. 65.];
 2  b = [18.; 11.; 3.; 73.];
```

Note that we are still in an environment with 6 worker processes, but the matrix A has only four rows, so our special precaution in the definition of mat_vec applies.

```
 3  n = length(b)
 4  x = zeros(n)                        # solution 0-initialized
 5  p = r = b
 6  rs_old = dot(r,r)                   # using dot
 7  for _ in 1 : n
 8      Ap = mat_vec(A,p)              # using mat_vec
 9      alpha = rs_old  / dot(p,Ap)    # using dot
10      global x += alpha*p
11      global r -= alpha*Ap
12      rs_new = dot(r,r)             # using dot
13      if sqrt(rs_new) < 1e-10 break end
14      global p = r + (rs_new / rs_old)*p
15      global rs_old = rs_new
16  end
17  println(x)   # ans: [1.0, 1.0, 1.0, 1.0]
```

13.7 Exercises

Exercise 13.1 Write a distributed version of the program to compute integrals in Exercise 8.1 in the Julia chapter.

Exercise 13.2 Write a distributed program for matrix multiplication.

Exercise 13.3 Based on the program in Example 6.8 in the C chapter, write a distributed recursive function that computes the determinant of regular matrices.

Exercise 13.4 In Sect. 8.9 in the Julia chapter, we developed a program to generate Julia set fractals. Write a distributed version of that program.

Exercise 13.5 Based on the program in Example 8.27 in the Julia chapter, write a distributed program to generate the colored version of the Julia set shown on the book cover. To do this, generate a 1000 × 1000 *shared matrix J* and distribute the computation of the matrix entries to the workers.

The result can then be visualized with the command `imshow(J, cmap = "hot")` from the package `PyPlot`.

Part V
Specialized Programming Environments

Chapter 14
Automated Solution of PDEs with FEniCS

The FEniCS Project is a steadily growing collaborative project for the development of innovative concepts and tools for automated scientific computing, with a particular focus on the automated solution of differential equations according to the so-called *finite element method FEM*.

The CS in FEniCS stands for Computer System, "ni" supposedly to provide a euphonious word. Possibly the name may also be an allusion to the University of Chicago mascot, a Phoenix rising from the ashes.

The FEniCS project began there in 2003, at the University of Chicago. It is currently being developed in collaboration with numerous international research groups.

FEniCS offers a collection of free software with an extensive list of features that can be embedded in the programming languages Python or C++.

Our discussion is based on Python.

Tools and Background

For an introduction and general information, the official site `fenicsproject.org` can be recommended, especially the e-book [6], which can be downloaded from `fenicsproject.org/tutorial` free of charge.

A good source for further studies is the 800-page book *Automated Solution of Differential Equations by the Finite Element Method* [7].

However, it should be mentioned that the FEniCS Project shows that it is still a "work in progress". The documentation is often rather sparse and incomplete. Even frequent version changes are not exactly helpful from a didactic point of view. For example, the introductory example in the FEniCS book 2012 is not compatible with the current FEniCS versions. The above tutorial is updated to version 1.6.2. But even here the introductory example does not run without changes. The changes may be minor, but can be a serious obstacle for the beginner.

The webpage `fenicsproject.org/download` contains instructions for free FEniCS installations for Linux, Windows and macOS, whereby the version for Ubuntu Linux may be the best choice with the least installation effort.

© Springer Nature Switzerland AG 2020
E. Smith, *Introduction to the Tools of Scientific Computing*, Texts in Computational Science and Engineering 25, https://doi.org/10.1007/978-3-030-60808-8_14

All examples in this chapter were tested under FEniCS 2019.1.0 based on Python version 3.6 in a virtual Ubuntu 18.04 environment on the macOS platform.

In Ubuntu, FEniCS can be installed with a few Terminal commands:

```
sudo apt-get install --no-install-recommends software-properties-common
sudo add-apt-repository ppa:fenics-packages/fenics
sudo apt-get update
sudo apt-get install --no-install-recommends fenics
sudo apt-get install python3-matplotlib
```

14.1 Finite Element Method, One-Dimensional

The finite element method FEM was developed mainly for the treatment of *partial* differential equations. The central ideas can however be illustrated with *ordinary* equations. In fact, we have already covered the basic principles in earlier chapters. In Sect. 5.6 in the SymPy chapter, we developed the *variational form* of such equations and showed how they can be solved with the Galerkin method.

In Sect. 10.7 in the Maple chapter, we followed up the development, and in Sect. 10.8 introduced special splines of degree 1, the so-called *hat functions*, as a basis for the Galerkin method.

And, put simply: Galerkin method with hat functions is at the core of the FEM.

A Simple Ordinary Differential Equation

For convenience, we briefly recall the main ideas.

We consider the following two-point boundary value problem with *homogeneous* Dirichlet conditions:

$$-u''(x) = f(x), \quad 0 < x < 1, \quad f(x) \equiv 8,$$
$$u(0) = u(1) = 0.$$
(1)

Variational Form

As explained in Sect. 5.6 in the SymPy chapter, a function u is a solution of $-u'' = f$ if and only if

$$-\int_0^1 u'' v = \int_0^1 fv$$
(2)

for all sufficiently smooth test functions v.

Galerkin Method

The general approach is now to fix a *finite* set of basis functions φ_i, and then try to establish equation (2), however replacing the exact goal function u_e with a function $u = \sum c_i \varphi_i$ (in FEniCS called *trial function*) built as a linear combination of the basis functions, with the coefficients c_i to be determined. This is the general Galerkin method.

Hat Functions

If we now choose hat functions as basis trial functions φ_i, we run into a difficulty, already discussed in Sect. 10.8 in the Maple chapter. For $u = \sum c_i \varphi_i$, the second derivative u'' referred to in (2) makes no sense, because hat functions do not have any sensible second derivative. Hat functions are only *piecewise differentiable*, without the derivatives being continuous. But in fact, continuity is not necessary. What makes all the difference for us here, is that the derivatives φ_i' are *integrable*, and therefore so are linear combinations with $u' = \sum c_i \varphi_i'$.

What we need is an equivalent formulation for (2), which however does not refer to *second* derivatives.

This is achieved through *integration by parts* of the left-hand side in (2). We get

(3)
$$-\int_0^1 u'' v = \int_0^1 u' v' - u' v \Big|_0^1.$$

Note that in (1) we assume homogeneous boundary conditions. Therefore we only have to consider test functions with $v(0) = v(1) = 0$, such that in particular the second term on the right-hand side in (3) vanishes. Hence we get

(4)
$$\int_0^1 u' v' = \int_0^1 f v$$

as an equivalent variational form for the differential equation $-u'' = f$. That is the one we will use in the following.

As mentioned in Sect. 5.6 in the SciPy chapter, it is customary to collect the left-hand side of the variational form in a bilinear form $a(u, v)$, and write the right-hand side, that does not depend on u, as a linear form $L(v)$.

Discretization

We formulate (4) for the special case that the trial function u is a linear combination $\sum_i c_i \varphi_i$ of the basis functions. By linearity of differentiation and integration, we then get the following approximate formulation for equation (4):

(5)
$$\sum_i c_i \int_0^1 \varphi_i' v' = \int_0^1 f v.$$

As a last step in the discretization we also restrict the test functions to the basis functions $v = \varphi_j$ themselves.

We come to the central equation on which our FEM is based:

Determine the coefficients c_i such that

(6)
$$\sum_i c_i \int_0^1 \varphi_i' \varphi_i' = \int_0^1 f \varphi_i \quad \text{for all } j.$$

Manual Solution

Before handing it over to the computer, however, we briefly illustrate by hand how the solution process works for our model example. To this end, we fix an equidistant sequence $x_0 = 0, x_1 = 0.25, x_2 = 0.5, x_3 = 0.75, x_4 = 1$. Let $\varphi_i, i = 1, ..., 3$, be the hat functions with $\varphi_i(x_i) = 1$, and $\varphi_i(x) = 0$ for $x \leq x_{i-1}$ and $x \geq x_{i+1}$.

A simple computation shows that

$$a_{ij} := \int_0^1 \varphi_i' \varphi_j' = \begin{cases} 2, & \text{if } i = j, \\ -1 & \text{if } |i - j| = 1, \\ 0 & \text{else.} \end{cases}$$

Similarly, recalling $f \equiv 8$ from (1), for all j:

$$b_j := \int_0^1 f \varphi_j = \int_0^1 8 \, \varphi_j = 2.$$

With $A = (a_{ij})_{1 \leq i, j \leq 3}$, $b := (b_j)_{1 \leq j \leq 3}^T$, and $c := (c_1, c_2, c_3)^T$ we thus get the equation system

(7)
$$Ac = b$$

with the solution $c = (0.75, 1, 0.75)$. Fig. 14.1 shows the approximation $u = \sum_i c_i \varphi_i$ represented by the solid line, the exact solution u_e by the dashed line.

Inhomogeneous Boundary Conditions

In our example we assumed homogeneous boundary conditions $u(0) = u(1) = 0$. However, this is no loss of generality, since the general case with $u(0) = \gamma_0, u(1) = \gamma_1$ can be easily reduced to the homogeneous one.

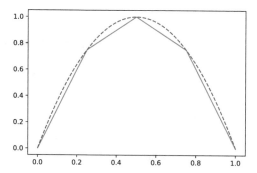

Fig. 14.1 FEM solution of equation (1) (solid line) and exact solution (dashed line)

To this end, instead of u we consider the function $\tilde{u} := u - r$, where $r(x) := (1 - x)\gamma_0 + x\gamma_1$, such that the linear equation system for $\tilde{u}$ again obtains the homogeneous form (1) above. The approximate solution $\sum_i \tilde{c}_i \varphi_i$ for $\tilde{u}$ is then immediately transformed into an approximate solution $\sum_i \tilde{c}_i \varphi_i + r$ for u.

14.2 FEniCS Implementation, One-Dimensional

We return to our model example (1) above, reformulated as a variational problem in (4) and discretized in (6). We solve the latter in FEniCS.

Assume the following code is stored in a file, say `onedim.py`.

FEniCS programs generally start with the import of the module `dolfin`, which contains the central classes and functions needed for FEniCS:

```
1 from dolfin import *
```

With the `dolfin` function `UnitIntervalMesh` we create a mesh for the unit interval, which divides it into 4 equal subintervals:

```
2 mesh = UnitIntervalMesh(4)
```

With `FunctionSpace` we then construct a space of functions over that mesh:

```
3 V = FunctionSpace(mesh, 'CG', 1)
```

The second argument in `FunctionSpace` specifies the type of the basis functions, the third argument the degree. Here we fix Continuous Galerkin functions of degree 1. Note that these are precisely our hat functions. As a synonym for 'CG' we can also write 'Lagrange'.

The following statements specify that our unknown solution function u (here declared as *trial function*, a term generally used in FEniCS) as well as all *test functions* v are to be constructed within this function space:

```
4 u = TrialFunction(V)
5 v = TestFunction(V)
```

Note that the spaces of trial and test functions will often coincide, though this is not compulsory.

Next comes the definition of the source function $f \equiv 8$ on the right-hand side of our equation (1):

```
6 f = 8
```

Now we need an expression for the boundary conditions:

```
7 bc = DirichletBC(V, 0, 'on_boundary')
```

In the present example, the constant value 0 is assigned to the whole boundary (which in our one-dimensional case consists precisely of the two boundary points 0 and 1 of the unit interval) with the FEniCS keyword 'on_boundary'.

Then the differential equation to be solved is formulated as a discrete variational problem:

```
8 a = u.dx(0)*v.dx(0)*dx
9 L = f*v*dx
```

The term 'u.dx(0)' denotes $\frac{\partial u}{\partial x}$ for the first argument (remember we are back in a programming environment where indices begin at 0). In our case, it simply means u', and analogously for v.

The notion '*dx' specifies that the usual integral over the whole interval is to be computed. Note that it has to be precisely 'dx' with the specific letter 'x'.

The variables a and L now represent the left-hand and right-hand side in the discrete variational form (6) above.

We come to the solution of the linear equation system $Ac = b$ in (7). For this we declare a function variable u_sol, to which the solution can be assigned:

```
10 u_sol = Function(V)
```

Then the discrete variational problem is solved by the dolfin function solve, and the result stored in u_sol:

```
11 solve(a == L, u_sol, bc)
```

Note that it is perfectly possible (and also customary) to use the same designator 'u' instead of u_sol.

A graphical representation of the solution can then be generated using the standard Python library Matplotlib.

The plot is prepared with the dolfin command

```
12 plot(u_sol)  # returns a matplotlib object
```

which can then be visualized with the matplotlib.pyplot function 'show':

```
13 from matplotlib.pyplot import show
14 show()
```

The program can be run by typing the following command in the Terminal window:

```
$ python3 "onedim.py"
```

Note that the command 'python3' specifically calls the Python version installed together with FEniCS.

Alternatively you can start the Python interpreter by

```
$ python3
```

and then at the Python prompt issue the command

```
>>> exec(open("onedim.py").read())
```

to execute the whole file content. Another possibility is to type the code line by line interactively.

Remark The program is executed as follows. The Python code refers to various modules written in C++. Before use, these must be compiled in machine code. This is achieved at runtime by a so-called Just in Time Compiler (JIT). This may lead to a slower execution when a program is run for the first time, but on successive runs the program can reuse the already compiled code.

Remark The solve function is the workhorse of FEniCS. Depending on the form of the input, it can apply various solution methods. Here the first argument is of the form 'a == L'. Therefore solve expects the left-hand side to be a bilinear form built from a TrialFunction and a TestFunction, and the right-hand side to be a linear form in the TestFunction. After the forms have been built in lines 8 and 9, in line 10 we could now redeclare 'u' as a Function to store the solution in line 11. However, for clarity, in this introductory example we refrain from doing so.

Remark 14.1 To get a better understanding of how FEniCS works, it may be useful to check the program step by step. To this end, directly after line 9 you can generate the matrix A and the vector b used in the solution process by

```
A = assemble(a); b = assemble(L)
```

and then represent them as NumPy arrays

```
A.array(); b.get_local()
```

Note that A is a dolfin *matrix*, b a *vector*, which, somewhat strangely, require different methods for the conversion to NumPy arrays.

Note also that for A we get a 5×5 matrix, and for b the vector 'array([1., 2., 2., 2., 1.])'.

The interested reader is invited to explain where the differences to our manual approach in Sect. 14.1 come from.

To see how the boundary conditions are worked into the equation, type

```
bc.apply(A,b)
```

and then check how that affects A and b.

To inspect the solution u_sol, we need it in dolfin vector form 'u_sol.vector()'. Then as above we apply the '.get_local()' method to return the coefficients as array([0., 0.75, 1., 0.75, 0.]).

14.3 Poisson Equations

As a model example for a *partial differential equation* PDE, we consider the following Poisson equation with *Dirichlet boundary conditions*, which is sometimes referred to as the "Hello World" of PDEs.

The general form of such an equation is

(1)
$$-\Delta u = f \quad \text{in a region } \Omega \subset \mathbb{R}^2,$$
$$u = g \quad \text{on the border } \partial\Omega,$$

where we here assume the special case $\Omega = [0, 1]^2$. As usual, Δu is an abbreviation for the sum of the second order partial derivatives $\dfrac{\partial^2 u}{\partial x^2} + \dfrac{\partial^2 u}{\partial y^2}$.

Method of Manufactured Solutions

Following the so-called *method of manufactured solutions*, we first construct a test problem, for which we can easily verify the solution. For this purpose we define the exact solution as
$$u_e(x, y) = 1 + x^2 + 2y^2.$$

Through backwards computation, we immediately obtain the source function f from the right-hand side in the Poisson equation (1) as
$$f = -\Delta u_e = -\Delta(1 + x^2 + 2y^2) = -(2 + 4) = -6.$$

On the boundary we use the exact solution $g := u_e$ itself.

Remark The method of manufactured solutions has the collateral benefit, that the highly non-trivial question, whether a differential equation is at all solvable, is here answered in the affirmative by definition.

Transformation into a Variational Problem

Analogous to the 1D case, (1) can equivalently be reformulated as a *variational problem*:

(2)
$$-\int_\Omega \Delta u \cdot v = \int_\Omega f \cdot v.$$

Again, similar to the 1D case, integrating the left hand side in (2) by parts allows an equivalent formulation in terms of first order derivatives, thus enabling the use of "less smooth" basis functions in the representation of the trial function u.

In full generality, the formula for integration by parts reads:

$$(3) \qquad -\int_\Omega \Delta u \cdot v = \int_\Omega \nabla u \cdot \nabla v - \int_{\partial\Omega} \frac{\partial u}{\partial \mathbf{n}} \cdot v.$$

Here, the rightmost term denotes the integral evaluation on the boundary $\partial\Omega$. However, as long as only Dirichlet boundary conditions are involved, we can also in the multidimensional case assume that all test functions v vanish on the boundary, such that the rightmost term will always be 0. We will return to the details later, in the context of Neumann boundary conditions.

For now, we concentrate on the constructions behind ∇u, and dually for v. '∇' (pronounced *nabla* or sometimes *del*) is an operator, that for a function, in our case a two-argument function $w \colon \mathbb{R}^2 \to \mathbb{R}$, denotes the vector valued *gradient*

$$\nabla w := \left(\frac{\partial w}{\partial x}, \frac{\partial w}{\partial y} \right) \colon \mathbb{R}^2 \to \mathbb{R}^2.$$

The product $\nabla u \cdot \nabla v$ is then to be understood as the scalar dot product between the two vectors, hence again resulting in a scalar-valued function.

To summarize: In the variational form of the problem (1) we are looking for a function $u \colon \Omega \to \mathbb{R}$, $u = g$ on $\partial\Omega$, such that

$$(4) \qquad \int_\Omega \nabla u \cdot \nabla v = \int_\Omega f \cdot v$$

for all sufficiently smooth test functions $v \colon \Omega \to \mathbb{R}$ with $v \equiv 0$ on the boundary $\partial\Omega$.

As mentioned earlier, it is customary to collect the left side of the variational form in a bilinear form $a(u, v)$ and write the right side, which does not depend on u, as a linear form $L(v)$.

Discretization and FEniCS Implementation

Similar to the 1D case, we approximate the continuous variational problem by a discrete one, assuming both the trial function u (expressing the solution) and the test functions v as linear combinations of suitable simple basis functions.

We again import the necessary tools from `dolfin` with

```
1 from dolfin import *
```

Regular Triangulation

Using the FEniCS function `UnitSquareMesh(n,m)`, we first construct a suitable mesh by *regular triangulation* of the region $\Omega = [0,1]^2$. More precisely, the unit square is divided into $n \times m$ rectangles of equal size, each of which is then partitioned into two triangles. Fig. 14.2 shows the construction for the 3×4 case.

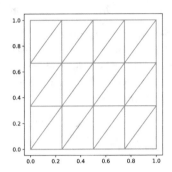

Fig. 14.2 3×4 triangulated unit square mesh

In the present case (and typical for two-dimensional partitions), these triangles are used to build the finite elements.

Here we choose

```
2 mesh = UnitSquareMesh(32,32)
```

Remark FEniCS provides many more useful basic classes for mesh generation, such as RectangleMesh in the 2D case, and UnitCubeMesh and BoxMesh for 3D problems.

Additionally, FEniCS offers powerful methods to construct more general mesh structures. We return to that later.

Discrete Function Space

We then define a function space over the mesh above, using

```
3 V = FunctionSpace(mesh, 'CG', 1)
```

Analogous to the hat functions in the 1D case, we assign a piecewise linear *pyramid function* $\varphi_i(x, y)$ to each inner node i of the mesh. The function assumes the value 1 at the node i, and vanishes at all other nodes. This is illustrated in Fig. 14.3.

The space V then consists of all linear combinations $v = \sum_i c_i \varphi_i$ of the basis functions.

Note that each $v \in V$ represents a locally linear function $a + bx + cy$ on each of the finite regions where it does not vanish.

Remark As alternatives to the continuous Galerkin functions of degree 1, FEniCS offers many more possibilities, for instance, spline functions built from piecewise polynomials of higher degree. A complete list can be obtained using the command help(FunctionSpace).

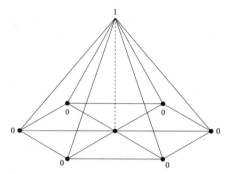

Fig. 14.3 Pyramid function

As in the 1D case, the variables for the solution function u and the test functions v are declared over the function space V:

```
4  u = TrialFunction(V)
5  v = TestFunction(V)
```

Source Function

From our manufactured solution we computed the source function to be $f \equiv -6$. In the program we can specify it by 'f=-6', as before, or as an instance of the Expression class by 'f=Expression('-6')'. For a constant however, the specification

```
6  f = Constant(-6.0)
```

is the most efficient, since it tells the underlying C++ compiler that the value does not have to be recomputed during program execution.

Boundary Condition

The function g that describes the boundary condition, is formulated according to C++ syntax, as an Expression object:

```
7  g = Expression('1 + x[0]*x[0] + 2*x[1]*x[1]', degree=2)
```

The degree specification serves to help FEniCS in the interpolation of the expression on the mesh, hence to increase computational precision. However in the present simple example there seems to be no need to bother.

The tool class Expression can also be used to construct more complex expressions, as can be seen from help(Expression).

The boundary condition itself is then again implemented as an instance of the class DirichletBC:

```
8  bc = DirichletBC(V, g, 'on_boundary')
```

The third argument `'on_boundary'` generates a discrete representation of the boundary $\partial\Omega$. Now `DirichletBC` specifies that the solution function u has to coincide with g on the boundary nodes.

The Discrete Variational Equation

As in the 1D case, the discrete solution is based on the construction of the matrix

$$A = (a_{ij}) \quad \text{with} \quad a_{ij} := \int_\Omega \nabla\varphi_i \cdot \nabla\varphi_j,$$

and the vector

$$b = (b_j) \quad \text{with} \quad b_j := \int_\Omega f \cdot \varphi_j,$$

into which then the boundary conditions are included during the actual solution process.

For the variational form we get

```
9  a = inner(grad(u), grad(v))*dx
10 L = f*v*dx
```

where `grad` denotes the gradient (cf. the '∇' operator above), `inner` the scalar dot product.

The computation of the solution then follows the same line as before:

```
11 u = Function(V)
12 solve(a == L, u, bc)
```

Note again that the boundary conditions are first incorporated into the equation during the solution process.

The solution can then be visualized graphically. As in the 1D case we use the `dolfin` function `plot` to prepare and hand over the solution to Matplotlib:

```
13 from matplotlib.pyplot import show
14 plot(u, mode="warp")
15 show()
```

The result is shown in Fig. 14.4.

14.4 Time-Dependent Poisson Equation

We consider a simple extension of the Poisson equation to a *time-dependent* equation that is often used to describe *heat dissipation* over time.

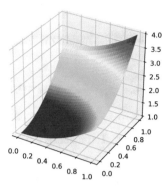

Fig. 14.4 FEniCS solution of PDE (1) with $f(x, y) = -6$, $g(x, y) = 1 + x^2 + 2y^2$

Our model has two spatial dimensions $s = (x, y) \in \Omega = [0, 1]^2$ and an additional time parameter $t \geq 0$:

$$
\begin{aligned}
\frac{\partial u}{\partial t}(s, t) - \Delta u(s, t) &= f(s, t), & s \in \Omega, & \quad t > 0, \\
u(s, t) &= g(s, t), & s \in \partial\Omega, & \quad t > 0, \\
u(s, 0) &= u_0(s), & s \in \Omega.
\end{aligned}
$$

(1)

Here f, g, u_0 are given functions. The solution $u = u(s, t)$, the source function $f = f(s, t)$ and the boundary condition $g = g(s, t), t > 0$, can change in space s and time t. The initial value u_0 depends only on space.

A Discrete-Time Approach

We define an equidistant sequence of time points $t_n := nh$ for a sufficiently small h, for which we inductively construct a sequence

$$u_n(s) \approx u(s, t_n)$$

of approximate solutions to the desired function u.

The start value u_0 is immediate from the third line in (1). Now assume u_n already to be defined. Then we determine u_{n+1} as the solution of the PDE

$$
\begin{aligned}
\frac{u_{n+1} - u_n}{h}(s) - \Delta u_{n+1}(s) &= f_{n+1}(s) := f(s, t_{n+1}), & s \in \Omega, \\
u_{n+1}(s) &= g(s, t_{n+1}), & s \in \partial\Omega.
\end{aligned}
$$

(2)

Here we use that for sufficiently small h, we can assume that

$$\frac{\partial u}{\partial t}(s, t_{n+1}) \approx \frac{u_{n+1} - u_n}{h}(s).$$

Transformation into a Variational Problem

By reordering the first line in (2), we obtain the following equation in the unknown function u_{n+1} on the left-hand side:

$$(3) \qquad u_{n+1} - h\,\Delta u_{n+1} = u_n + h f_{n+1}.$$

The variational problem we have to solve thus becomes:

Find $u_{n+1}: \Omega \to \mathbb{R}$, such that for all sufficiently smooth $v: \Omega \to \mathbb{R}$ with $v \equiv 0$ on $\partial\Omega$:

$$\int_\Omega u_{n+1}\, v - \int_\Omega h\,\Delta u_{n+1}\, v = \int_\Omega u_n\, v + h f_{n+1}\, v.$$

Integration by parts, observing the boundary condition for v, yields

$$\int_\Omega \Delta u_{n+1} v\, ds = -\int_\Omega \nabla u_{n+1} \cdot \nabla v\, ds,$$

such that the variational problem can be formulated as:

Determine u_{n+1}, such that for all v:

$$(4) \qquad \int_\Omega u_{n+1}\, v + h\,\nabla u_{n+1} \cdot \nabla v = \int_\Omega u_n\, v + h f_{n+1} v.$$

Manufactured Solution

As in the stationary case, we construct a test problem for which we can easily check the solution.

We first fix the exact solution

$$u_e = 1 + x^2 + 3y^2 + 1.2t,$$

from which we also immediately get

$$(5) \qquad u_0 = 1 + x^2 + 3y^2.$$

Inserting u_e into the first line of the heat equation (1) yields

$$f = \frac{\partial u}{\partial t} - \Delta u = 1.2 - 2 - 2 \cdot 3 = -6.8.$$

Note that then also all functions f_n, $n > 0$, assume the same constant value -6.8.

The boundary condition is once again determined by the exact solution

$$(6) \qquad g = u_e = 1 + x^2 + 3y^2 + 1.2t.$$

FEniCS Implementation

We can now formulate a FEniCS program that, for a given time point T, approximates $u(s, T)$ with a function $u_T(s)$. Observe that here u_T depends only on space, i.e. the time parameter does not appear as a variable.

We first fix initial and last time points 0 and 1, as well as the discrete time steps $h = 0.1$ for the temporal process:

```
1 t = 0.; h = 0.1; T = 1.0
```

As usual we need:

```
2 from dolfin import *
```

Mesh and function space are again introduced by

```
3 mesh = UnitSquareMesh(32,32)
4 V = FunctionSpace(mesh, 'CG', 1)
```

The variables u_sol and v to store the solution functions u_n and test functions v, are then declared over this function space V:

```
5 u_sol = Function(V)   # not TrialFunction!
6 v = TestFunction(V)
```

Note that u_sol is declared as a general `Function`, not as `TrialFunction`. The reason is explained in the remark below.

The right-hand side of the PDE (1), hence also the value of all f_n in (4), is given by the constant expression computed above:

```
7 f = Constant(-6.8)
```

To encode the boundary condition according to (6) we first define a Python string

```
8 g_str = '1 + x[0]*x[0] + 3.0*x[1]*x[1] + 1.2*t'
```

We use g_str to instantiate an `Expression` object to specify the boundary conditions. Note however that only expressions with explicit numerical values for the extra parameter t are accepted by the constructor `DirichletBC`.

As a start value for $t = 0$ we get

```
9 g = Expression(g_str, t=0, degree=2)
```

For the start-value u_0 of our function sequence according to (5) we can use the same expression g. However we first have to convert it to a function in the discrete function space V.

This is achieved through

```
10 u_n = project(g,V)   # initial value u0 according to (5)
```

In the following loop, the functions u_n are determined iteratively according to (4):

```
11  while t <= T:
12      a = u_sol*v*dx + h*inner(grad(u_sol), grad(v))*dx
13      L = u_n*v*dx + h*f*v*dx
14      g.t = t  # assign new t-value to boundary expression
15      bc = DirichletBC(V, g, 'on_boundary')
16      solve(a - L == 0, u_sol, bc)  # nonlinear solver
17      u_n.assign(u_sol)            # next in u_n sequence
18      t += h                       # next time step
```

Lines 12 and 13 represent the left- and right-hand sides in (4). The variable u_n holds the current last sequence entry u_n. The successor entry u_{n+1} is to be stored in u_sol.

But note that a is not a legal bilinear form accepted by the solver, because u_sol is not declared as TrialFunction. In such cases, it is necessary to pass the equation in the 'F == 0' form with F as 'a - L'. Again, see the remark below for details.

The assignment g.t = t in line 14 gives the expression g the new t value, required to update the boundary condition bc in line 15.

Line 16 performs the actual solution.

In line 17 the result is assigned to the variable u_n by the method assign, such that in the next iteration round it is available as the new last sequence entry.

The final result for $T = 1$ can then be plotted. Note, however, that the plot does not show the complete temporal evolvement.

Remark The variable u_sol plays a double role, on the one hand in the *specification* of the variational form in line 12, on the other hand in the *solution* to store the result in line 16. For the latter it must be of type Function, *not* TrialFunction. But a variable of type Function is not accepted by the solver in an argument of the form 'a == L'.

Here the 'F == 0' form comes to the rescue, since the solver can deal with a general Function variable when it appears in an argument of *that* form. The disadvantage is that the solver then uses a more complex *nonlinear* solution method.

14.5 Nonlinear Equations

We illustrate how to solve *nonlinear* differential equations in FEniCS.

Consider the ordinary differential equation

$$-w' = 0 \quad \text{in } [0, 1],$$

(1) $$\text{where} \quad w = (1 + u)^2 \cdot u',$$

$$u(0) = 0, \ u(1) = 1,$$

with the exact solution $u_e = (7x + 1)^{1/3} - 1$. The term $(1 + u)^2$ makes it nonlinear in the unknown function u.

Variational Form

As a variational problem we get: Determine u, such that

$$-\int_0^1 w' \cdot v = \int_0^1 0 \cdot v = 0$$

for all sufficiently smooth v with $v(0) = v(1) = 0$.

By the usual partial integration of the left hand side, observing the second line in (1), the form to be solved becomes

(2)
$$\int_0^1 (1+u)^2 \cdot u' \cdot v' = 0.$$

Picard Iteration

We solve (2) with the so-called *Picard iteration*. The idea is to inductively define a sequence of functions $u_0, u_1, \ldots, u_n$, such that in every step, u_{n+1} is chosen as the solution u of the following variational form, now indeed *linear* in u:

(3)
$$\int_0^1 q(u_n) \cdot u' \cdot v' = 0, \quad q(u_n) := (1 + u_n)^2,$$

where u_0 should be chosen close to the real solution.

We develop a corresponding solution program.

The FEniCS Program

The program begins as usual:

```
1 from dolfin import *
2 mesh = UnitIntervalMesh(10)
3 V = FunctionSpace(mesh, 'CG', 1)
4 v = TestFunction(V)
```

However, we postpone the declaration of a variable to store the solution function. The reason is that we want to reuse the present code block in the next program below.

We introduce the nonlinear factor as:

```
5 def q(u): return (1 + u)**2
```

We come to the boundary conditions:

```
6 def lb(x): return near(x[0], 0)
7 def rb(x): return near(x[0], 1)
8 lc = DirichletBC(V, 0, lb)
9 rc = DirichletBC(V, 1, rb)
10 bc = [lc, rc]
```

Here the dolfin function near in lines 6 and 7 denotes 'indiscernible within machine precision'. Line 6 essentially specifies that a floating point number x belongs to the left boundary, if it is indistinguishable from 0. Dually, line 7 specifies a point to belong to the right boundary, if it is indistinguishable from 1.

In lines 8 and 9 the two boundary conditions are defined, and then collected into a single condition in line 10.

We come to the definition of the Picard iteration.

The variable u_n is used to inductively store the functions u_n in (3), beginning with the initial value $u_0 \equiv 0$:

```
11  u_n = project(Constant(0), V)   # u_n is of type 'Function'
```

We now need a function variable 'u', to which the solver can assign the new iterative solution u_{n+1}, depending on the previous u_n, currently stored in u_n:

```
12  u = Function(V)   # not TrialFunction
```

The iterative computation of the approximate solutions is to be repeated until

$$||u_{n+1} - u_n||_\infty < \epsilon := 10^{-6},$$

where we however abort the process after 25 steps to avoid non-terminating computations. To compute the supremum norm we use the NumPy function 'norm' with the option 'ord=inf'. We get:

```
13  eps = 1.0e-6
14  from numpy import inf; from numpy.linalg import norm
15  for _ in range(25):
16      F = (q(u_n)*u.dx(0)*v.dx(0))*dx
17      solve(F == 0, u, bc)
18      diff = u.vector().get_local() - u_n.vector().get_local()
19      d = norm(diff, ord=inf)
20      if d < eps: break
21      u_n.assign(u)
```

The variational problem (3) is formulated in line 16. In line 17 it is passed to the solver in 'F == 0' form, where a variable of type Function is accepted.

In line 18 the difference between new and old approximation is computed. Observe that the suffix 'vector().get_local()' denotes the method that assigns the values of the FEniCS Function object to a NumPy array.

In line 19 the maximum of the value differences is assigned to the variable 'd' through the NumPy function norm, using the maximum option.

In line 20 the iterative computation is terminated, when the desired precision is reached. Otherwise, the result is assigned to u_n, and a new iteration round is initiated.

After the end of the loop, the final result is stored in u. As usual, it can be plotted using the dolfin function plot to pass the result to Matplotlib.

Direct Solution

Actually, FEniCS already contains ready-made tools for the automatic solution of nonlinear differential equations.

We illustrate the approach for the equation (1) above.

The first part of the program is the same as in the Picard iteration in the last example. The code block consisting of lines 1–10 can be copied literally.

We continue with the new line 11 and declare a variable 'u' of type TrialFunction, which we then use to build the variational form:

```
11  u = TrialFunction(V)
12  F = inner(q(u)*grad(u), grad(v))*dx
```

Remark Note that the variational form is defined in a way that immediately extends to the multidimensional case. In the present one-dimensional case, the expression 'inner(grad(u),grad(v))' is automatically evaluated as 'u.dx(0)*v.dx(0)'.

Once the variational form has been built, we can redeclare 'u' as a Function variable to represent the solution:

```
13  u = Function(V)
```

The solution algorithm itself is in the present case based on the Newton method (instead of a Picard iteration). Therefore, we need the derivative of the expression F in line 12.

First, the command 'action(F,u)' transforms F into a discrete *vector* of the function values of u on the mesh nodes, from which then the derivative is generated in form of a Jacobian matrix J:

```
14  F = action(F,u)
15  J = derivative(F,u)
```

The solution process still needs a start value for u. In general, this choice may be critical, but here again, the following is sufficient:

```
16  u_init = project(Constant(0), V)
17  u.assign(u_init)
```

Based on these preparations, we can now specify the problem as an instance of a suitable problem class:

```
18  problem = NonlinearVariationalProblem(F, u, bc, J)
```

and instantiate a solver object specifically adapted to such non-linear problems:

```
19  solver = NonlinearVariationalSolver(problem)
```

and finally apply a specific solve method of that solver object:

```
20  solver.solve()   # assigns solution to u
```

The solution u can then be plotted as in the Picard-version above.

14.6 Neumann Boundary Conditions

So far we have only considered boundary conditions of the Dirichlet type, where the *value* of the unknown function u is specified on the boundary.

In contrast, a *Neumann boundary condition* consists in specifying the *derivative* of the solution, using a function g:

$$\frac{\partial u(x)}{\partial \mathbf{n}} = g(x),$$

where $\mathbf{n}(x)$ denotes the perpendicular unit vector that points outward from the boundary $\partial\Omega$ in the boundary point x.

If, as we will do in the following, we limit ourselves to PDEs over the 2D domain $\Omega = [0, 1]^2$, then the directional derivatives can be obtained directly from the partial derivatives $\dfrac{\partial u}{\partial x}$ and $\dfrac{\partial u}{\partial y}$.

Mixed Dirichlet-Neumann Boundary Conditions

We first give an example, in which both Dirichlet and Neumann conditions occur.

So, let $\Omega = [0, 1]^2$. Let $\Gamma_D = \{(x, y) \in \partial\Omega \mid x = 0 \text{ or } x = 1\}$ be the vertical component of the boundary $\partial\Omega$, and $\Gamma_N := \partial\Omega - \Gamma_D$ the horizontal part, where however the corner points belong to Γ_D.

Consider the Poisson equation

$$
\begin{aligned}
-\Delta u &= f &&\text{in} && \Omega = [0, 1]^2, \\
u &= g_D &&\text{on} && \Gamma_D \ \ (\text{Dirichlet condition}), \\
\nabla u \cdot \mathbf{n} &= g_N &&\text{on} && \Gamma_N \ \ (\text{Neumann condition}).
\end{aligned}
$$

(1)

Manufactured Solution

By the method of manufactured solution, we again choose $u = 1 + x^2 + 2y^2$ as the exact solution, and adjust f, g_D, and g_N accordingly.

We get $f \equiv -6$ and $g_D = 1 + x^2 + 2y^2$ as before.

For the Neumann condition, we define the function $g_N(x, y) = 4y$, that yields the values $g_N(x, 0) = 0$ and $g_N(x, 1) = 4$ on Γ_N.

Variational Form

For processing with FEniCS, the PDE must again be formulated as a variation problem.

In Sect. 14.3 we developed the general variational form

(2)
$$\int_\Omega \nabla u \cdot \nabla v = \int_\Omega fv + \int_{\partial\Omega} \frac{\partial u}{\partial \mathbf{n}} \cdot v,$$

and noted that as long as we only considered Dirichlet conditions, we could limit ourselves to test functions that vanished on the boundary, which is why the second term on the right-hand side could always be assumed to be 0.

This time we cannot omit the boundary term arising from the integration by parts. The test functions v can only be assumed to be 0 on Γ_D.

Hence our model example in variational form becomes:

Determine u, such that, observing the Dirichlet condition above

(3)
$$\int_\Omega \nabla u \cdot \nabla v = \int_\Omega f \cdot v + \int_{\Gamma_N} g \cdot v$$

for all sufficiently smooth test functions v with $v = 0$ on Γ_D.

We collect the left side again in a bilinear form $a(u, v)$, and the now modified right side in $L(v)$.

Here we note a fundamental difference between Dirichlet- and Neumann conditions: The Neumann condition appears *explicitly* in the formulation of the variational problem itself. A Dirichlet condition, on the other hand, is not considered until it is woven into the linear equation system in the solution process.

FEniCS Implementation

We first note that in FEniCS, a Neumann condition automatically applies precisely to that part of the boundary, where no Dirichlet conditions are specified.

The FEniCS implementation begins as usual:

```
1 from dolfin import *
2 mesh = UnitSquareMesh(32,32)
3 V = FunctionSpace(mesh, 'CG', 1)
4 u = TrialFunction(V)
5 v = TestFunction(V)
```

The boundary region Γ_D for the Dirichlet condition can be defined as follows, again with the dolfin function 'near':

```
6 def dir_boundary(x): return near(x[0], 0) or near(x[0], 1)
```

For this boundary region we can now specify the Dirichlet condition:

```
7 g_D = Expression('1 + x[0]*x[0] + 2*x[1]*x[1]', degree=2)
8 bc = DirichletBC(V, g_D, dir_boundary)
```

The discretization of the variational problem:

```
9  f = Constant(-6)
10 g_N = Expression('4*x[1]', degree=1)
11 a = inner(grad(u), grad(v))*dx
12 L = f*v*dx + g_N*v*ds   # note 'ds' with s
```

In line 12 it is important to note that FEniCS interprets the operator '*dx' as a command to evaluate the integral on the whole of the the domain, whereas the operator '*ds' specifies that it is to be computed along the boundary.

Note that the integration '*ds' is carried out over the entire boundary, including the Dirichlet boundary. However, since the test functions v vanish on the Dirichlet boundary (as a consequence of specifying a DirichletBC object), the integral will only include the contribution from the Neumann boundary.

The solution:

```
13 u = Function(V)
14 solve(a == L, u, bc)
```

Plotting the result shows that the solution coincides with the one in Sect. 14.3.

Pure Neumann Boundary Conditions

We consider an equation, where all boundary conditions are of the Neumann type. Note that such a problem is well posed, only if a function value for u in non-differentiated from is specified somewhere in the equation, since specifications that only refer to derivatives can only determine a solution up to an additive constant.

We make the ansatz to include u itself into the equation, like this:

$$-\Delta u + u = f \quad \text{in } \Omega = [0,1]^2,$$
(4)
$$\nabla u \cdot \mathbf{n} = g \quad \text{on } \partial\Omega.$$

Manufactured Solution

We consider the manufactured solution

$$u(x, y) = \cos(\pi x)\cos(\pi y).$$

As can easily be seen (or established with SymPy or Maple), we get

(5) $$f(x, y) = (1 + 2\pi^2)\cos(\pi x)\cos(\pi y), \quad g(x, y) \equiv 0.$$

Variational Form

The variational form has the same shape as in (3) above, except that the term '$+uv$' is added to the left-hand side, whereas on the right-hand side the term referring to the integral over the boundary is not needed, since it is 0 by the assumption $g \equiv 0$.

The variational form thus reads: Determine u, such that

(6)
$$\int_\Omega \nabla u \cdot \nabla v + u\,v = \int_\Omega f \cdot v$$

for all sufficiently smooth test functions v.

FEniCS Implementation

The FEniCS program is now straightforward.

As usual it begins with:

```
1 from dolfin import *
2 mesh = UnitSquareMesh(32,32)
3 V = FunctionSpace(mesh, 'CG', 1)
4 u = TrialFunction(V)
5 v = TestFunction(V)
```

The bilinear form a, corresponding to the left-hand side in (6):

```
6 a = inner(grad(u), grad(v))*dx + u*v*dx
```

For the linear form L on the right-hand side we set

```
7 f = Expression('(1 + 2*pi*pi)*cos(pi*x[0])*cos(pi*x[1])', degree=1)
8 L = f*v*dx
```

We come to the solution:

```
9  u = Function(V)
10 solve(a == L, u)  # no reference to boundary conditions
```

which can then be plotted as usual:

```
11 from matplotlib.pyplot import show
12 plot(u, mode="warp")
13 show()
```

The plot is shown in Fig. 14.5 below.

14.7 Stokes Equation

So far we have considered PDEs in a *single, scalar* valued unknown function. We now turn to a *system*, consisting of two functions, a *vector valued* function u, and a scalar valued p.

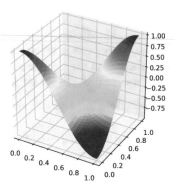

Fig. 14.5 FEniCS solution of equation (4) above with f, g as in (5)

More precisely, we consider the following simple steady-state variant of the so-called *Stokes equation*:

(1)
$$\begin{aligned}
-\Delta u + \nabla p &= f \quad \text{in } \Omega := [0,1]^2, \\
\operatorname{div} u &= 0 \quad \text{in } \Omega, \\
u &= g \quad \text{on } \partial\Omega,
\end{aligned}$$

where $u\colon \Omega \to \mathbb{R}^2$ can be interpreted as the velocity field of an incompressible fluid motion, and $p\colon \Omega \to \mathbb{R}$ is then the associated pressure. $f, g\colon \Omega \to \mathbb{R}^2$ are source terms.

Mathematical Preliminaries

Let $u = (u_1, u_2)\colon \mathbb{R}^2 \to \mathbb{R}^2$ be a vector valued function.

The Laplace operator is then applied componentwise, resulting in a vector valued function
$$\Delta u = (\Delta u_1, \Delta u_2)\colon \mathbb{R}^2 \to \mathbb{R}^2.$$

The *divergence* $\operatorname{div} u$ of $u = u(x_1, x_2)$ is the scalar valued function given by

$$\operatorname{div} u := \frac{\partial u_1}{\partial x_1} + \frac{\partial u_2}{\partial x_2}.$$

Below, we will need the *gradient* operator ∇ for vector valued functions. It is also applied componentwise, such that $\nabla u = (\nabla u_1, \nabla u_2)\colon \mathbb{R}^2 \to \mathbb{R}^4$ returns a 2×2 matrix, since both components ∇u_i are vector valued.

In various previous examples, integration by parts for scalar functions u, v led to the scalar dot product $\nabla u \cdot \nabla v$. In the present case, with vector valued $u = (u_1, u_2)$ and $v = (v_1, v_2)$, that role is played by the so-called *double dot product*, extending the dot product to a scalar multiplication between the 2×2 matrices ∇u and ∇v.

It is defined as follows:

$$\nabla u : \nabla v := \sum_i \sum_j \frac{\partial u_i}{\partial x_j} \cdot \frac{\partial v_i}{\partial x_j}.$$

Variational Problem

In transforming (1) to a variational problem, a useful approach is to consider a *mixed* formulation.

We start with the equation in the first line in (1). Let V be a space of sufficiently smooth test functions $v : \Omega \to \mathbb{R}^2$ with $v \equiv 0$ on $\partial\Omega$.

As usual we multiply the equation by a test function v, and then take the integral over Ω, reducing terms with second derivatives to first order expressions through integration by parts.

For the leftmost term $-\Delta u$, we get

$$(2) \qquad -\int_\Omega \Delta u \cdot v = \int_\Omega \nabla u : \nabla v,$$

and, similarly, for the term ∇p:

$$(3) \qquad \int_\Omega \nabla p \cdot v = -\int_\Omega (\operatorname{div} v)\, p.$$

Now, to the second line in (1). Here, we again use a set Q of sufficiently smooth functions $q : \Omega \to \mathbb{R}$ as test functions. Then the equation written in variational form becomes

$$(4) \qquad \int_\Omega (\operatorname{div} u)\, q = 0.$$

Adding the equations above, the variational problem to be solved reads:

Given f, g and the boundary conditions $u = g$ in the third line in (1) above, find functions u, p, such that:

$$(5) \quad \int_\Omega \nabla u : \nabla v - \int_\Omega (\operatorname{div} v)\, p + \int_\Omega (\operatorname{div} u)\, q = \int_\Omega f \cdot v \quad \text{for all } v \in V,\ q \in Q.$$

FEniCS Implementation

By the method of manufactured solutions, we again first construct a test problem, for which we can easily verify the solution.

To this end we define the exact solution by

(6)
$$u_e(x, y) := \begin{pmatrix} \cos(\pi y) \\ \sin(\pi x) \end{pmatrix},$$

$$p_e(x, y) := \pi \cos(\pi x) \cos(\pi y).$$

Note that $\int_\Omega p_e = 0$. This is easily verified using e.g. SymPy or Maple.

Through backwards-computation, the function f from the right-hand side of the equation (1) then becomes

(7)
$$f(x, y) = \begin{pmatrix} \pi^2 \left(\cos(\pi y) - \sin(\pi x) \cos(\pi y) \right) \\ \pi^2 \left(\sin(\pi x) - \sin(\pi y) \cos(\pi x) \right) \end{pmatrix}.$$

On the boundary $\partial\Omega$ we again use the exact solution $g := u_e$ itself.

The Program

The FEniCS program begins as usual:

```
1 from dolfin import *
2 mesh = UnitSquareMesh(32,32)
```

To specify our mixed test space, corresponding to the Cartesian product $W := V \times Q$, in the current dolfin version we must first declare two types of elements:

```
3 ue = VectorElement('CG', triangle, 2)
4 pe = FiniteElement('CG', triangle, 1)
```

Note that for the VectorElement we demand a degree 2. The reason is that the Stokes equation is known to be a *saddle point* system. For such systems the pyramid functions are in general not smooth enough to guarantee a stable solution.

We combine the element types to a "mixed" type 'ue*pe', and, based on the compound elements, define a discrete function space W over the mesh:

```
5 W = FunctionSpace(mesh, ue*pe)
```

This mixed finite element space W is known as the *Taylor–Hood elements*, and is a stable standard element space for the Stokes equations.

In W we can then declare variables for the trial and test functions, given as pairs of the original element types:

```
6 (u,p) = TrialFunctions(W)
7 (v,q) = TestFunctions(W)
```

We come to the specification of the exact solution for u in (6), needed in the boundary condition:

```
8 u_ex = Expression(('cos(pi*x[1])', 'sin(pi*x[0])'), degree=1)
```

The boundary condition is specified only for the velocity function u, hence we must extract the corresponding component W.sub(0) from the function space W:

```
9 bc = DirichletBC(W.sub(0), u_ex, 'on_boundary')
```

For the variational form we need the function f from (7):

```
10  f = Expression(('pi*pi*(cos(pi*x[1]) - sin(pi*x[0])*cos(pi*x[1]))',
11                   'pi*pi*(sin(pi*x[0]) - sin(pi*x[1])*cos(pi*x[0]))'),
12                   degree=1)
```

The variational form is then a straightforward implementation of (5). Note that FE-niCS automatically adapts the 'inner' and 'grad' functions to the vector valued situation:

```
13  a = inner(grad(u), grad(v))*dx - p*div(v)*dx + q*div(u)*dx
14  L = inner(f,v)*dx
```

We can now solve the discrete variational problem, incorporating the boundary condition in line 9:

```
15  w = Function(W)
16  solve(a == L, w, bc)
17  (u,p) = w.split(True)
```

In line 16 the solution consisting of the computed function *pair* is assigned to the variable w. The Function method `split` then splits the compound result into the desired two goal functions.

A detailed explanation of the exact effect of the `split` method, including the significance of the parameter `True` can be inquired with 'help(Functions)'.

Visualizing

We come to the graphical representation of the solution, again based on Matplotlib:

```
18  from matplotlib.pyplot import show, figure, colorbar
```

First, we consider the velocity u:

```
19  figure()
20  v = plot(u)
21  colorbar(v)
```

In contrast, for the pressure p the matter is somewhat more complicated. Note that in the equation system (1), the function p appears only in derived form, which implies that the solution will be unique only up to an additive constant. A standard way to circumvent this, is to include a *normalizing* specification into the equation system, such as $\int_\Omega p = 0$. Recall that this condition holds for our manufactured solution above.

   In the present case, the normalization can be achieved as follows: Let c be the array of coefficients c_i of the basis functions φ_i in the representation of the solution function p found by FEniCS. If we then define a new function $p_{\text{norm}} := \sum_i (c_i - c_{\text{avg}})\varphi_i$, where c_{avg} is the average over all c_i, then this function has the desired properties. Within computational precision, it is a solution of the equation system (1), where now also $\int_\Omega p_{\text{norm}} = 0$.

In FEniCS, the following instruction converts the original solution to such a p_{norm}:

```
22 normalize(p.vector())
```

We can now generate the pressure graph:

```
23 figure()
24 plot(p, mode='warp')
```

and then display both plots:

```
25 show()
```

Fig. 14.6 shows the solution for the vector field u and the scalar pressure p.

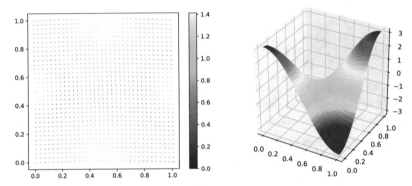

Fig. 14.6 FEniCS solution for the Stokes equation (1) with f as in (7), $g = u_e$ in (6), at the left the velocity field u, at the right the pressure p

14.8 Adaptive Mesh Refinement

So far we always assumed a fixed mesh resolution, e.g. `UnitSquareMesh(32,32)`. It is however obvious that better approximations can be obtained by considering finer meshes. In order to avoid unnecessary computing effort, it is recommended to refine a mesh only in regions in which the unknown function is expected to vary widely.

Such an adaptive refinement is easy to implement in FEniCS. We illustrate the idea with an example.

For this purpose we consider the 2D Poisson equation

$$
-\Delta u = f := 100\, e^{-100((x-0.2)^2+(y-0.3)^2)} \quad \text{in } \Omega = [0,1]^2,
$$
(1)
$$
u = 0 \quad \text{on } \partial\Omega.
$$

The core method is to construct a sequence of meshes G_n and approximate solutions u_n with respect to G_n until the solutions differ only by a given tolerance, for example

(2) $$|M(u_{n+1}) - M(u_n))| \leq tol$$

for a measure function M, e.g. $M(u) := \int_\Omega u$, and a tolerance level $tol > 0$.

The solver function `solve` can include the additional arguments tol and M, iterating the solution process on increasingly finer meshes until (2) holds.

The main observation is that `solve` only refines the mesh G_n *locally* around those nodes, that promise the largest precision gain.

FEniCS Implementation

As customary we start with

```
1 from dolfin import *
```

For illustration we choose a mesh, that is obviously too coarse:

```
2 mesh = UnitSquareMesh(4,4)
```

The next declarations should again be self explanatory:

```
3  V = FunctionSpace(mesh, 'CG', 1)
4  bc = DirichletBC(V, 0, 'on_boundary')
5  u = TrialFunction(V)
6  v = TestFunction(V)
7  f = Expression('100*exp(-100*(pow(x[0]-0.5, 2) + pow(x[1]-0.3, 2)))',
8                  degree=1)
9  a = inner(grad(u), grad(v))*dx
10 L = f*v*dx
11 u = Function(V)
```

We define the measure function M and the tolerance level tol, as explained above:

```
12 M = u*dx
13 tol = 1.e-3  # for illustration only - in practice e.g. 1.e-6 better
```

The adaptive solution process is then started with these values, which are inserted in the argument placeholders in `solve`:

```
14 solve(a == L, u, bc, tol=tol, M=M)
```

The plots in Fig. 14.7 below show the solutions with regards to initial and final mesh as well as the meshes themselves:

```
15 from matplotlib.pyplot import show, figure
16 figure(); plot(u.root_node(), mode='warp')
17 figure(); plot(u.leaf_node(), mode='warp')
18 figure(); plot(mesh)
19 figure(); plot(mesh.leaf_node())  # FEniCS 1.7.2
20 show()
```

Remark Note that line 19 was executed in FEniCS version 1.7.2. The generation function 'mesh.leaf_node()' it is no longer available in later versions.

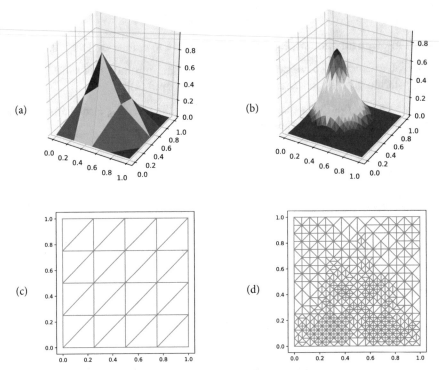

Fig. 14.7 Adaptive mesh refinement. (a) shows solution on initial coarse mesh, (b) the one on the mesh adaptively refined by the FEniCS program, (c) and (d) the underlying meshes

14.9 User Defined Meshes

So far we have considered meshes from the basic FEniCS collection. It is however obvious that in modeling of real world problems, in general rather more complex structures are required. In the following we briefly discuss some useful methods to generate meshes.

The Package mshr

The package mshr provides tools to construct various domains based on simple geometric shapes such as Circle or Rectangle. These can then be combined to more complex shapes, using the Boolean operations intersection, union, and difference. The resulting regions can be superposed with different kinds of meshes, using the operator generate_mesh.

As an example, we solve a PDE over an *annulus*. In mathematics, an annulus (from the Latin word "anulus" for "little ring") is a ring shaped object, more precisely a region bounded by two concentric circles.

Mesh Generation

We consider an annulus Ω, where the outer circle has a radius of 1, the inner one a radius of 0.5.

Using tools from mshr, we first provide a FEniCS representation of the domain Ω:

```
1 from dolfin import *
2 from mshr import *
3 c1 = Circle(Point(0.0), 1)
4 c0 = Circle(Point(0.0), .5)
5 annulus = c1 - c0
```

On this domain we define a discrete mesh with

```
6 mesh = generate_mesh(annulus, 20)
```

The result is shown in Fig. 14.8 (for illustration with a coarser node density).

```
7 from matplotlib.pyplot import show
8 plot(mesh)
9 show()
```

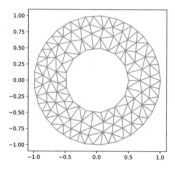

Fig. 14.8 Mesh generated on annulus

The mesh can then used in FEniCS programs like any prefabricated mesh. We'll see an example below.

For later use we store the mesh in an xml-file:

```
10 file = File('annulus.xml')
11 file << mesh  # note: output operator as in C++
```

Within the file, the *nodes* are represented like this:

```
<vertex index="42" x="-1.913417161825452e-01"
                   y="-4.619397662556433e-01" />
```

and the *triangles* like this:

```
<triangle index="381" v0="42" v1="299" v2="449" />
```

Application Example

We illustrate the use in a program to solve the Laplace equation

$$\Delta u = 0 \quad \text{in } \Omega,$$

(1) $u = \sin(5x) + \cos(5y)$ on the outer boundary (Dirichlet condition),

$\nabla u \cdot \mathbf{n} = 0$ on the inner boundary (Neumann condition).

The following program is as desired:

```
1 from dolfin import *
```

The mesh is imported from the file `annulus.xml`:

```
2 mesh = Mesh('annulus.xml')
```

Remark Alternatively we could just as well continue the mesh-generation program above, and append the following program part directly after line 6.

Then a discrete function space is induced on this mesh:

```
3 V = FunctionSpace(mesh, 'CG', 1)
```

We define the Dirichlet condition for the outer boundary:

```
4 g_out = Expression('sin(5*x[0]) + cos(5*x[1])', degree=1)
5 def out_boundary(x, on_boundary):
6     return (x[0]**2 + x[1]**2 > .8 ) and on_boundary
7 bc = DirichletBC(V, g_out, out_boundary)
```

To recall: when for some part of the boundary there is no Dirichlet condition specified, then automatically a Neumann condition is assumed to hold there. If none is specified, then one with the default value 0 is assumed. And that is exactly what we need for the inner boundary in the present case.

The encoding of the variational problem then reads:

```
8  f = Constant('0.0')
9  u = TrialFunction(V)
10 v = TestFunction(V)
11 a = inner(grad(u), grad(v))*dx
12 L = f*v*dx   # Neumann condition is 0
```

The solution:

```
13 u = Function(V)
14 solve(a == L, u, bc)
```

The result is shown in Fig. 14.9:

```
15 from matplotlib.pyplot import show
16 plot(u, mode='warp')
17 show()
```

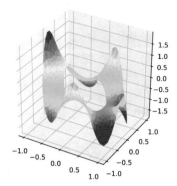

Fig. 14.9 FEniCS solution of equation (1) over annulus mesh

Manual Mesh Refinement

In Sect. 14.8 we saw how adaptive mesh refinement can be performed automatically
by the FEniCS system. The techniques can however also be applied directly by the
user. The following FEniCS program illustrates the idea:

```
1 from dolfin import *
2 from mshr import *
```

Consider a domain made from a rectangle from which a circular segment has been
subtracted:

```
3 dom = Rectangle(Point(0,0),Point(1.2, .4)) - Circle(Point(.3,.2), .15)
```

We first construct a coarse mesh:

```
4 mesh = generate_mesh(dom, 8)
```

which we then intend to refine, for illustration first for the whole mesh, and then
restricted to the left half of the rectangle.

The refinement for the whole mesh is obtained by

```
5 mesh_fine = refine(mesh)
```

To refine only the left part, we need to indicate the triangles involved by specifying
markers:

```
6 markers = MeshFunction('bool', mesh, 2)  # 2 = mesh-dimensions
7 markers.set_all(False)  # initialized
```

Then the triangles to be refined are specified:

```
8 for c in cells(mesh):
9     if (c.midpoint()[0] < .6): markers[c] = True
```

Each of the marked triangles is then refined. This results in the mesh

```
10 mesh_partial_fine = refine(mesh, markers)
```

We plot all three meshes:

```
11  from matplotlib.pyplot import show, figure
12  figure(); plot(mesh)
13  figure(); plot(mesh_fine)
14  figure(); plot(mesh_partial_fine)
15  show()
```

Fig. 14.10 shows the comparison.

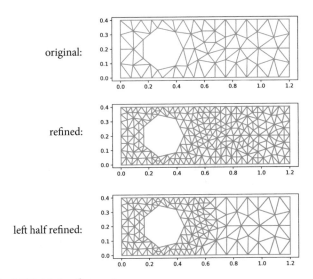

Fig. 14.10 Mesh refinement

Mesh Libraries

Actually, there is already a wealth of meshes freely available for many standard situations, but also for rather more exotic ones. A good collection can, for instance, be downloaded from the site `fenicsproject.org`.

When it comes to large real-world projects, however, the built in mesh-construction tools from FEniCS become quite cumbersome. Then it is more sensible to resort to other specialized tools, and import the resulting meshes back into FEniCS.

One such tool that is often used in connection with FEniCS is Gmsh, which can be downloaded for free from the site `gmsh.info`.

14.10 Final Note: Parallel Processing in FEniCS

We have seen that a main component of the finite element method is the solution of matrix equations. This naturally raises the question of whether FEniCS can benefit from distributing such calculations over several processes.

In fact, we can run any FEniCS code in parallel using MPI as explained in the Python Chapt. 11, simply by initiating the computation for multiple instances of the Python interpreter.

Example 14.2 Assume the code in Sect. 14.3 to be stored in a file poisson.py.
The computation can then be distributed to four processes by

```
$ mpiexec -n 4 python3 poisson.py
```

This will return 4 partial results, including separate plots for each.

At the moment, however, there does not seem to be an obvious elementary method to combine the partial results and diagrams into a common global solution.

14.11 Exercises

First Order ODEs

Exercise 14.1 Write FEniCS programs to solve the following differential equations:

(1)
$$u'(x) = 2x \ \text{ in } [0, 1], \quad u(0) = 0,$$

(2)
$$u'(x) = \frac{u(x)}{1 + x^2} \ \text{ in } [0, 1], \quad u(0) = 1.$$

Second Order ODEs

Exercise 14.2 Consider the ODE

$$-u'' + u = x \ \text{ in } [0, 1].$$

(1) Compute the solution assuming Dirichlet boundary conditions

$$u(0) = u(1) = 0.$$

For control: The exact solution is $u_e(x) = x - \sinh(x)/\sinh(1)$.

(2) Now replace the right-hand side condition by the *Neumann condition*

$$u'(1) = 1 - \cosh(1)/\sinh(1),$$

and write a solution program. The exact solution is the same as in case (1).

Exercise 14.3 Solve the ODE

$$-\frac{d}{dx}\left((1+x)\frac{du}{dx}\right) = x^2 \text{ in } [0,1],$$

$$u(0) = u(1) = 0.$$

Partial Differential Equations PDEs

Exercise 14.4 Solve the Poisson equation

$$\Delta u = x^2 + y^2 \qquad\qquad \text{in } [0,2] \times [0,1],$$
$$u(x,y) = x(2-x) + y(1-y) \quad \text{on the boundary.}$$

Observe that the domain is is a rectangle.

Exercise 14.5 Consider once more the PDE

$$(*) \qquad \begin{aligned} -\Delta u &= f \equiv -6 &&\text{in } \Omega = [0,1]^2, \\ u &= 1 + x^2 + 2y^2 &&\text{on the boundary } \partial\Omega. \end{aligned}$$

(1) Embed the solution program into a function, that includes the mesh size n as argument.

(2) Extend the program to one, which additionally estimates the error, by computing the difference to the exact solution with respect to the L^2 integral norm. For this, use the dolfin function errornorm. For details consult help(errornorm).

(3) Solve the problem for $n = 10, 20, 30$.

Exercise 14.6 Solve the same PDE ($*$), now however using *quadratic spline functions* instead of hat functions.

Exercise 14.7 Solve ($*$), now however considering the 3D domain $\Omega = [0,1]^3$ instead of the 2D version $[0,1]^2$. For testing, choose a new "manufactured solution" and adapt the program accordingly.

Exercise 14.8 We consider the heat equation in Sect. 14.4. For the computation of the iterative solution, we used a "backward Euler" method, which determined the current iterative solution u_n *implicitly* from the equation

$$u_{n+1} - h\Delta u_{n+1} = u_n + h f_{n+1}.$$

(1) Change the code, such that it now uses the direct forward Euler method

$$u_{n+1} = u_n + h\Delta u_n + h f_n.$$

What happens?

(2) Back to the original formulation, change the manufactured solution $u = 1 + x^2 + 3y^2 + 1.2t$, such that the source function f for the right-hand side becomes time-dependent. Adapt the solution program accordingly.

The Third Mode of solve

For the time-dependent Poisson equation in Sect. 14.4, we applied the FEniCS solver solve with a *nonlinear* solution method (in line 16 of the program). However, solve can operate in yet another mode: a linear system $Ax = b$ may be solved by calling solve(A,x,b), where A is a matrix and x and b are vectors.

Exercise 14.9 Modify the solution program accordingly. To this end, declare the function u_sol to be of type TrialFunction, generate the bilinear form 'a' and the corresponding matrix A, using the techniques in Remark 14.1 in Sect. 14.2. Then redeclare u_sol as a Function object. This is done outside of the while loop.

Inside the loop, construct the vector 'b', work the boundary condition into the equation, pass it to solve, where the solution is inserted into the function u_sol in vector form u_sol.vector().

A Neumann Challenge

The following exercise should mainly be considered as an inspiration.

In Sect. 14.6, we discussed a variant of our standard Poisson equation with mixed Dirichlet and Neumann boundary conditions.

Here the task is to develop an essentially pure Neumann version. A viable procedure seems to be, based on the same manufactured solution as before, to define Dirichlet conditions for one or more of the corner points of $\Omega = [0, 1]^2$, then appropriate Neumann conditions g_N for the remaining boundary.

It will probably be difficult to find a closed expression for g_N. As an alternative, subclassing of the Expression class by overriding its eval method, could permit to specify g_N using normal Python definitions.

As a further possibility, the boundary could be partitioned into separate subdomains.

Exercise 14.10 Develop a variant of the Poisson equation in Sect. 14.6, based as far as possible on pure Neumann boundary conditions.

References

1. T. A. Driscoll, R. J. Brown, *Fundamentals of Numerical Computation.* (SIAM, 2017). Julia versions of the source code can be downloaded from `github.com/tobydriscoll/fnc-extras/tree/master/julia`
2. R. Johansson, *Numerical Python: Scientific Computing and Data Science Applications with Numpy, SciPy and Matplotlib.* (Springer, Berlin, 1996)
3. B. W. Kernighan, D. M Ritchie, *The C Programming Language.* (Prentice Hall, 1988)
4. W. Gander, M. Gander, J. Martin, F. Kwok, *Scientific Computing - An Introduction using Maple and MATLAB.* (Springer, Berlin, 2014)
5. H. P. Langtangen, *A Primer on Scientific Programming with Python.* (Springer, Berlin, 2016)
6. H. P. Langtangen, A. Logg, *Solving PDEs in Python: The FEniCS Tutorial I.* (Springer, Berlin, 2017)
7. A. Logg, K. A. Mardal, G. Wells (eds.), *Automated Solution of Differential Equations by the Finite Element Method: The FEniCS Book.* (Springer, Berlin, 2012)
8. M. Lutz, *Learning Python.* (O'Reilly Media, 2013)
9. The Mathwork Company, *MATLAB Primer.* (2019), retrieved from `www.mathworks.com/help/pdf_doc/matlab/getstart.pdf`
10. A. Quarteroni, F. Saleri, G. Paola, *Scientific Computing with MATLAB and Octave.* (Springer, Berlin, 2014)
11. J. R. Shewchuk, *An Introduction to the Conjugate Gradient Method Without the Agonizing Pain.* (1994), retrieved from `www.cs.cmu.edu/~quake-papers`
12. E. Smith, *Elementare Berechenbarkeitstheorie.* (Springer, Berlin, 1996)
13. E. Smith, *Carl Adam Petri. Life and Science.* (Springer, Berlin, 2015)
14. B. Stroustrup, *The C++ Programming Language.* (Addison-Wesley, 2013)

© Springer Nature Switzerland AG 2020

E. Smith, *Introduction to the Tools of Scientific Computing*, Texts in Computational Science and Engineering 25, https://doi.org/10.1007/978-3-030-60808-8

Index

The index is divided into individual sections. The first list contains general subjects and persons. Programming-specific terms are grouped by language in order of appearence: Python, C/C ++, Julia, MATLAB, Maple.

Subjects and Persons

A

Abel, Niels Henrik   105
Ackermann function   37, 120, 155
Anaconda, Python distribution   20
Angular velocity   77
Anonymous function   35, 155, 198
Assignment operator   10

B

Bezanson, Jeff   149
Binary word   14
Boundary oscillation   235
Boundary value problem
        77, 78, 91, 108, 109, 147,
        173, 204
Bratu equation   80, 174, 205
Butterfly effect   2
BVP   *see* Boundary value problem

C

Characteristic polynomial   65

Chebyshev nodes   234
Cholesky decomposition   62
Church's thesis   12
Church, Alonzo   12
Class   42
Collatz problem
        28, 29, 119, 152, 153, 215
Column-major order   55, 168, 196
Compiler   16, 115
Conditional statement
        28, 118, 152, 192
Conjugate gradient
        58, 59, 166, 254, 263, 274, 285
Conjugate vectors   58
Continued fraction   179
Cramer's rule   2

D

Dalcin, Lisandro   240
Deadlock   242
Diophantine equation   14
Dirichlet boundary condition   81

© Springer Nature Switzerland AG 2020
E. Smith, *Introduction to the Tools of Scientific Computing*, Texts in Computational
Science and Engineering 25, https://doi.org/10.1007/978-3-030-60808-8

Python

C/C++

Julia

Symbols

Matlab

Maple